Water Bioengineering Techniques
for Watercourse Bank and Shoreline Protection

Also of interest

Ground Bioengineering Techniques
for Slope Protection and Erosion Control
H.M. Schiechtl and R. Stern
0–632–04061–0

The Rivers Handbook: Two Volume Set
Edited by P. Calow and G.E. Petts
0–86542–848–4

River Restoration
Edited by G.E. Petts and P. Calow
0–86542–919–7

Water Bioengineering Techniques

for Watercourse, Bank and Shoreline Protection

H.M. Schiechtl

and

R. Stern

Translated by L. Jaklitsch
UK Editor David H. Barker

b

**Blackwell
Science**

© 1994 Österreichischer Agrarverlag,
Klosterneuburg, Austria
© 1997 English translation with additions
Blackwell Science Ltd

Editorial Offices:
Osney Mead, Oxford OX2 0EL
25 John Street, London WC1N 2BL
23 Ainslie Place, Edinburgh EH3 6AJ
238 Main Street, Cambridge
  Massachusetts 02142, USA
54 University Street, Carlton
  Victoria 3053, Australia

Other Editorial Offices:
Arnette Blackwell SA
  224, Boulevard Saint Germain
  75007 Paris, France

Blackwell Wissenschafts-Verlag GmbH
  Kurfürstendamm 57
  10707 Berlin, Germany

  Zehetnergasse 6
  A-1140 Wien
  Austria

First published in German as *Handbuch für naturnahen
Wasserbau* by Österreichischer Agrarverlag, Austria
First published in English by
Blackwell Science 1997

Set in 10 on 13 pt Times
by DP Photosetting, Aylesbury, Bucks
Printed and bound in Great Britain
by Hartnolls Ltd, Bodmin, Cornwall

The Blackwell Science logo is a
trade mark of Blackwell Science Ltd,
registered at the United Kingdom
Trade Marks Registry

DISTRIBUTORS

  Marston Book Services Ltd
  PO Box 269
  Abingdon
  Oxon OX14 4YN
  (*Orders:* Tel: 01235 465500
         Fax: 01235 465555)

USA
  Blackwell Science, Inc.
  238 Main Street
  Cambridge, MA 02142
  (*Orders:* Tel: 800 215-1000
         617 876-7000
    Fax: 617 492-5263)

Canada
  Copp Clark Professional
  200 Adelaide Street, West, 3rd Floor
  Toronto, Ontario M5H 1W7
  (*Orders:* Tel: 416 597-1616
         800 815-9417
    Fax: 416 597-1617)

Australia
  Blackwell Science Pty Ltd
  54 University Street
  Carlton, Victoria 3053
  (*Orders:* Tel: 03 9347-0300
    Fax: 03 9347-5001)

A catalogue record for this title
is available from the British Library

ISBN 0–632–04066–1

Library of Congress
Cataloging-in-Publication Data

Schiechtl, Hugo M.
    [Handbuch für naturnahen Wasserbau. English]
    Water bioengineering techniques: for watercourse
bank and shoreline protection/H.M. Schiechtl and
R. Stern; translated by L. Jaklitsch; UK editor,
David H. Barker.
       p. cm.
    Includes bibliographical references and index.
    ISBN 0–632–04066–1 (alk. paper)
    1. Streambank planting. 2. Shorelines.
3. Bioengineering.
I. Stern, R. II. Title.
TC537.S3513 1997
627′.58–dc20   96-38475
                                                   CIP

Contents

List of Colour Plates

Preface

During the past few years, the attitude of society to the environment has experienced a significant change. Many people in all walks of life are becoming aware of the progressive impoverishment of the landscape and the accompanying degradation of the environment taking place in many parts of the world.

For these reasons, working with nature generally and nature-oriented restoration and reclamation measures are bound to be increasingly considered as options. It points the way to greater emphasis on bio-technical methods in civil engineering and the implementation of successful practical measures for structured erosion control and earth stabilisation.

The aim of this handbook is to bring the benefits of using vegetative construction or bioengineering techniques of riverbank and shore protection to the notice of as wide a range of bioengineering professionals engaged in practical fieldwork as possible. This should stimulate the interest of and encourage engineers and others to employ such bioengineering measures whenever interference with the natural landscape in the course of development causes problems requiring rectification.

At the same time, attention is drawn to the ever-increasing volume of relevant publications on the subject, and to the need for closer exchange of ideas and co-operation between experts engaged in this field.

In order to foster the use of bioengineering methods, it is important, and desirable, that intensive training on them is given to those already working on the practical aspects of erosion control and protection. The latest information on successful bioengineering applications should also be made available to others with an interest in securing the maximum useful life of riverbanks and shorelines. Unfortunately, the reliability of available mathematical slope stability models incorporating vegetative effects is hampered by the general lack of reliable input data, but this is not normally a major aspect of water bioengineering design.

Editor's Note

This handbook complements the publication *Ground Bioengineering Techniques for Slope Protection and Erosion Control* by the same authors. It provides a rare opportunity to gain insights into the approach of the chief exponent, Professor Hugo Schiechtl, and his colleague, Dr Roland Stern, of the discipline of water bioengineering: the use of vegetation for the engineering and ecological enhancement of riverbanks and shores. The book has been written after a lifetime engaged in developing for modern situations the classical repertoire of earlier bioengineers working in the Central Alpine region of Europe. The techniques described in detail in Chapter 3 have been built upon a profound knowledge of local soils, plants, their ecology and fluvial processes. This has provided a foundation for the protection and stabilisation of riverbanks and shorelines in the natural and formed landscape. As such, the book deals primarily with shallower non-navigable waterways and lakes, but its contents have potential for wider application. The vegetation-based techniques for the protection and stabilisation of the fringes of waterbodies range from wholly vegetative 'soft' techniques to 'semi-hard' or composite structures such as stone, gabions or timber crib work with vegetative inclusions. These are treated in three main groups: soil protection techniques, ground stabilisation techniques and combined construction techniques, the latter subdivided into transverse and longitudinal structures.

There is also a chapter on bioengineering techniques in earth dam and floodbank construction. This challenges conventional thinking of the British school of water engineers, and perhaps those of many other countries, which maintain that there is no place for vegetation along banks of watercourses and canals and on crests and slopes of floodbanks and earth dams. The bioengineering approach to water engineering has been carried out in Central Europe for centuries, and used to be carried out until comparatively recently in the UK with satisfactory results. Changing management practices in recent decades have threatened the effectiveness of shrub- and tree-based bank protection and stabilisation systems.

Though cost savings have been claimed as the reason for the reduction in bank vegetation maintenance, in reality the consequential costs in terms of the degradation in condition of watercourse banks, the remedial works need to counter them and damage to river corridor habitats have been as much if not greater. In addition, realisation has dawned that 'soft' engineering techniques can be as effective as 'hard' ones. Though often needing nurturing after completion, pending full establishment, and regular maintenance thereafter, bioengineered installations have much lower initial or capital costs, resulting in lower whole-life costs.

This handbook is especially timely as there has been increasing recognition in the English-speaking world of the contribution of European workers to the development of waterway and shoreline engineering with emphasis on conservation and sustainable practice. The publication of this book coincides also with increasing official and public awareness that there has been excessive intervention in natural drainage systems resulting in unsatisfactory hydrological performance of catchments owing to drainage of wetlands, encroachment into flood plains by housing and other developments, and channelisation of watercourses, often using hard lining systems. This has been accompanied by loss of habitats and resulting damage to wildlife, fisheries, recreation and visual amenity.

The information in *Water Bioengineering Techniques for Watercourse Bank and Shoreline Protection* focuses inevitably on vegetative techniques for protection of lake shores and banks of rapidly flowing watercourses typical of the authors' working base, the Central European Alpine region. These techniques are likely to cater for the most erosive situations occurring along any river, including those along spillways and below weirs in lower reaches of rivers.

As such the book complements *The New Rivers and Wildlife Handbook*, published by the Royal Society for the Protection of Birds, the National Rivers Authority and the Royal Society for Nature Conservation (Ward *et al.*, 1994). This deals with lowland situations primarily, reflecting in turn the topography of the UK. It covers river processes, biology, survey methods and management practices to benefit wildlife, with only a short section on engineering techniques involving the use of vegetation, though there is a full and varied section on case studies, again mainly selected from streams and rivers in lowland areas. Flood defence and land drainage are also treated in the holistic manner of the rest of the book and in harmony with the work of the European school of water bioengineering as detailed herein.

An interesting aspect of *The New Rivers and Wildlife Handbook* is that it derives from an independent homegrown strand of river engineering, with inputs from the latest UK-based research into fluvial processes and river biology, having no direct connection with the work of European water bioengineers. The UK's approach can only be strengthened and invigorated by the exposure to the wide spectrum of mainland European bioengineering techniques set out by Professor Schiechtl and Dr Stern in this handbook. It is hoped that similar benefits will apply to other regions in the world where past nature-hostile approaches are beginning to give way to more balanced ecological engineering strategies. Readers are urged to adopt (and adapt for local conditions) the sustainable tactics and techniques described in this book to aid this process of harnessing rather than harming Nature.

Synonyms

In general there are many different terms describing similar techniques involving the use of vegetation for civil engineering purposes in the four main active German-speaking countries or regions – Austria, Germany, Switzerland and the Alto Adige-South Tirol region in Northern Iealy. A difficult problem is thus compounded for non-German speakers wishing to decipher original texts, increasing the value of this book.

As the practice of vegetative engineering has begun to spread around the world, the most commonly adopted English term to describe it is bioengineering, a direct translation of the most commonly used German-language term 'Ingenieurbiologie'. However, the disciplines of human and genetic engineering are both frequently called bioengineering and/or biotechnology. Since the medical and genetic use of the prefix 'bio' in conjunction with the words engineering or technical pre-dated and has a much higher 'profile' than vegetative engineering topics, the shared use of the terms bioengineering without suitable prefix or 'identifying' link words can give rise to distracting confusion. This is made more complicated as genetic engineering of plants offers the prospect of improving the civil or geotechnical engineering performance of some plants to enhance their desirable characteristics.

Since the term 'soil bioengineering' has been defined by Sotir (1995) as involving plants exclusively, it is considered that the best prefix to use is 'ground', as in 'ground bioengineering'. This mirrors the associated term 'ground engineering' which is defined as 'concerning engineering processes for improving ground'. Ecological engineering or eco-engineering are other terms that have been proposed (Nordin, 1993, 1995). Also 'biotechnical slope protection' is used frequently by bioengineering practitioners in the USA to describe the combined use of structural and vegetative elements to arrest and prevent slope failures and erosion (Gray, 1991).

Notes to the text

Mention is made frequently in the text to woody plants capable of throwing shoots and roots from live cuttings and branches; for the Central European Alpine region, these are invariably shrub willows, selected for their hardiness, vigour and non-invasive qualities, especially the purple willow (*Salix purpurea*). Dogwood (*Cornus sanguinea*) and in favourable circumstances most hardy shrubs can also be established from cuttings.

Figures are not drawn to scale.

Acknowledgements

The assistance of Dr Neil G. Bayfield on ecological aspects of the text and Mrs Klare Ware on the translation from the German of references and further reading items and communications with the authors is gratefully acknowledged. Thanks also go to Julia Burden at Blackwell Science for her faith in the text, the discipline of water bioengineering and the editor.

David H. Barker
Managing Director, Geostructures Consulting, Edenbridge

Introduction

Water, the basic requirement of all life, resists any force/deviation from the path of least resistance. The banks of all natural watercourses and the entire catchment areas are, under normal and non-arid conditions, covered with vegetation. It would therefore be very shortsighted to ignore the ecological relationships that exist in any given river system and to exclude the use of vegetative means from enhancing the accepted classical engineering methods employed in river training projects and other water engineering works.

Watercourses and the vegetation types associated with them are an integral part of the landscape. To preserve, maintain and, where necessary, re-shape these systems to keep pace with the demands of development are challenging tasks for the planner and the implementing agency.

The aim of this book is to outline the basic role that bioengineering techniques can play in the planning and construction of water resource protection measures, while also considering the need for maintenance and the cost of such works during the ensuing years. This is not the place to deal with the basics of hydrology and hydraulics or geology and soil mechanics. These subjects are barely touched on since there are many excellent publications available on these topics.

A companion to *Ground Bioengineering Techniques for Slope Protection and Erosion Control*, this handbook looks at the surface- and shallow-seated stabilisation of banks and waterways and river courses, including the range of available construction materials and specific construction methods. Its main aim is to stimulate interest in, and encourage the use of, live plant material in the construction of durable erosion control measures. This will bring together the often divergent interests of those concerned with carrying out and maintaining the works, those concerned with the protection of natural resources and the environment, landscape architects, adjoining property owners, and, last but not least, the ecologist.

Chapter 1
Planning and Implementation

Live building materials have been used for centuries for the protection of river and stream banks, lakes and sea shores. During the Middle Ages when neither modern building materials nor construction machinery were in use, shores and riverbanks were successfully stabilised and protected by the use of plants and plant materials; in France and the Netherlands, even navigable canals were kept in good repair using the same methods. Many of these conservation methods and measures were, however, gradually forgotten only to be rediscovered after the turn of the last century.

Knowledge gained from recent research into biological and technical aspects and the properties of building materials contributed substantially to the improvement of these ancient building methods and pointed the way to new areas of application.

Today, the civil engineer has a wide range of bioengineering methods at his disposal which, combined together with the correct choice of natural building materials available on site, make construction methods using such natural materials feasible and possible. Ever-increasing population pressure has made flood protection one of the most important subjects for water or hydraulic engineers: flood protection measures have thus held centre stage during the past 200 years or so. The use of more and larger machinery, in conjunction with preferred 'hard' construction methods and the hunger for more land, consequently led to the loss of wetland and water retention areas and with it to the loss of many plant species in both limnetic and terrestrial plant associations.

Greater awareness of the state of our environment appears to have firmly taken root, and is clearly demonstrated by society's initiative in demanding a more natural approach to the conservation of water resources and natural wetland areas.

1.1 Planning projects to fit into the landscape

It should be noted that the use of water bioengineering techniques for the protection and stabilisation of banks of rivers and canals and edges of other water bodies, is in most cases only part of the water engineer's task. The optimum use of natural methods and materials in the construction of protection works is only feasible if the envisaged measures fit harmoniously into the landscape. If the technical aspects of a project design do not make proper allowance for such methods they will be relegated to a purely aesthetic function.

This concept must be fully understood in order to appreciate the role that water bioengineering techniques can play within the total range of civil protection measures.

It is not the intention of the authors to dwell on the basic principles of water resource planning and protection, but to highlight the more important aspects of a natural approach to these disciplines following the principles laid down by Kauch (1992) and the Water Board of the Austrian Ministry of Agriculture and Forestry (1991).

The use of natural building materials requires space. It would be futile to attempt the implementing of vegetative methods in the construction of protection measures if the necessary space is not available.

One of the main tasks in river regulation works is the safe disposal of flood waters; this requires structures that can withstand the attendant forces. The hydraulic and technical objects are achieved cost effectively by integrating live building materials in the required structures.

Every landscape is characterised by its own typical drainage pattern according to its geomorphological characteristics. In intensively cultivated and settled areas, the original reticulation system of rivers and streams has largely been replaced by artificial channels and river courses. The term 'natural' therefore can only apply to some remnant lengths of our rivers, not to mention still waters and lakes.

The following criteria apply in general to water conservation and protection measures that incorporate biological or ecological considerations:

❑ Retention and maintenance of the natural drainage pattern.
❑ Retention and maintenance of the river dynamics.
❑ Retention and maintenance of pools, riffles and point bars.
❑ Retention and maintenance of the original bed level.
❑ Retention and maintenance of shorelines and riverbanks.
❑ Retention and maintenance of the variable width of the natural watercourse.

❏ Retention and maintenance of the variable and different flow patterns.
❏ Preservation of water retention areas.
❏ Retention and maintenance of migratory routes of water organisms and aquatic life.
❏ Use of live building and construction materials.
❏ Establishment, development and maintenance of natural vegetation adapted to the project site(s).
❏ Sustained maintenance-oriented economic use of riparian and seasonally inundated areas according to sound silvicultural practices.
❏ Emergency plan in case of structural failure and damage limitation.
❏ Due consideration for the multi-functional nature of the watercourse or resource in settled areas.
❏ Due consideration for the economic advantage of employing natural materials and live plants in the construction of protection and erosion control measures, and allowance in the project plan for eventual repair work after flood damage.

The regulation, protection and stabilisation of river courses, particularly if these works incorporate natural methods, should fit harmoniously into the landscape. Before the final decision is made as to which bioengineering technique is to be employed, the alignment of the watercourse and the design of the banks, shoreline, etc. must be determined, including longitudinal and transverse sections.

The natural alignment of any river course in its lower reaches consists of a series of regular left- and right-hand bends or meanders with variable radii. This basic feature was described by Fargue as long ago as 1867. He pointed out that the width of the bed should increase from a point at the start of the bend towards its apex (Lange and Lecher, 1989). The most natural alignment is that which corresponds as closely as possible to the type of watercourse to be dealt with. A sequence of artificial bends in white waters or in upland torrents is as unnatural as the straightening of a riverbed in its lower reaches by confining the flow to an artificial channel between bends.

The difference between minimum flow and high flood is often so great that a single river channel cannot cope adequately with both conditions. In this situation a narrow meandering low water channel can be constructed within a broad and straight flood watercourse or river corridor. A natural riverbed is made up of pools, shallows and deep sections; the gradient will vary along its course and with it the speed of flow. Pools fulfil many functions: they provide a valuable habitat for aquatic life and reduce speed of flow, thus diminishing the forces that would otherwise expend their energy on riverbank protection works; they also regulate to a certain degree the movement of the base load. The deepest pools are usually confined to the narrowest bends in the river course which leads to reduced flow velocities. Sections of fast river flow or riffles coincide with the transition between two opposite bends. This should be taken into consideration when planning new bed alignments in order to avoid uniform velocities throughout.

As there are no 'standard' riverbed profiles in nature, such regimented approaches should equally be avoided when employing natural or bioengineering techniques in riverbed regulation or protection. If water bodies are to provide a suitable habitat for aquatic life, sections of variable width and depth are essential. Great variability in the design of cross-sections is not only of advantage for stability reasons, but it also lends itself, for ecological and aesthetic reasons, to the efficient application of vegetative construction methods.

The choice of transverse section type has a direct bearing on the choice of building material, be it inert sand, stone and cement and/or living plants. Structures for the protection of riverbanks and shorelines very often involve a certain amount of earthworks, which necessitate the involvement of several technical disciplines at the planning stage. The civil engineer in many cases will call upon the services of the geologist, botanist, biologist and the expert in soil mechanics. Detailed investigations on site, supplemented by laboratory analysis, will provide in broad outline the concept of the cross-sections suited to the prevalent terrain conditions. Erosion, bed material, bank gradients below and above the average water level, and stability of both cut and watercourse bank sections will receive due consideration.

1.2 Implementing projects in harmony with the natural landscape

Considerations and guidelines include the following:

❏ Careful selection of suitable construction machinery and tools matched to terrain characteristics.
❏ Stable and correctly shaped watercourse banks; avoidance of steep gradients.
❏ Use of local building materials, e.g. stone, gravel, sand, soil, wood.
❏ Use of local building materials that do not naturally occur at the construction site, e.g. rocks and boulders in fine grained alluvials, is best avoided.
❏ Avoidance of artificial building materials, e.g. steel, concrete, plastics for surface cladding or grouting of river or stream beds.

- ❏ Preferential use of live building materials.
- ❏ Obtaining woody plants capable of vegetative propagation from the construction site, its environs or from similar nearby habitats.
- ❏ Preservation of reed beds and aquatic plants within the river regulation area.
- ❏ Preservation of vegetation on the fringes of the construction or regulation area by the considerate use of earth moving machinery and equipment.
- ❏ Removal, temporary storage and re-establishment (transplantation) of vegetation.
- ❏ Restricted or, at best, total avoidance of cutting of traces, fragmentation or clearing of alluvial woodland.

Erosion forces, soil and pore water pressure, tractive forces and artesian water pressure are largely concentrated and superimposed at the riverbed/riverbank interface. Inadequate protection in this area may lead to undercutting. As a result, the riverbank will be eroded up to the median water level, which may lead to the total collapse of the bank. The correct shaping and adequate erosion control of the lower riverbank section is therefore a prerequisite for the successful protection of the upper sections.

It is frequently unavoidable that hard construction methods have to be used in conjunction with vegetative methods. The use of hard materials should, however, be limited or confined to those sections where purely vegetative means are inadequate to prevent erosions (Anselm, 1976). This is the case when one or several of the following points apply:

- ❏ The tractive force and flow velocity exceed the resistance of the bed and bank material.
- ❏ Newly established vegetative measures need protection until full rooting has taken place.
- ❏ Increasing ground water pressure displaces fine grained materials of the bed and lower bank.
- ❏ There is insufficient space for vegetative measures, usually the case within built-up areas.

1.3 Planning of water bioengineering construction

Due to the lack of technical training and experience, there is still a certain reluctance to resort to bioengineering techniques and stabilisation methods. In many instances, such methods are not part of the original project plan, but resorted to when the usual hard structures show signs of failure. Remedial measures based on the use of vegetative means are then hastily implemented, without proper planning and preparation. It is therefore advisable, at the preliminary planning stage, to seek the advice of an engineer trained in the use of bioengineering techniques to establish the areas where such means may be integrated into the traditional civil engineering works based on the use of solid construction methods. A certain amount of time is required to arrive at the most suitable and economic solution to the problem. The final decision on the selection of the most suitable vegetative materials and the construction method can only be made after site investigations have ascertained the local conditions, and due consideration has been given to the wishes of the client with regard to the final appearance of the project site.

The checklist in Table 1.1 will aid in arriving at a decision as to which points are crucial for the preparation of the plan, and which are of lesser or of no importance. The decision on what bioengineering technique to use and the choice of live plant materials will largely influence the programme for the project implementation.

Work on site will start, in the first instance, in those areas that need

Table 1.1 Checklist for the planning of water bioengineering construction.

No.	Type of work
1	Obtain topographical maps, aerial photos, orthophotos* and construction plans
2	Site visit, alignment route, longitudinal and cross-sections, hydrological information
3	Evaluation of geological and hydrogeological investigations
4	Evaluation of the soil analysis results of the bed materials and watercourse bank stability
5	Evaluation of hydrographic data
6	Evaluation of the vegetation survey of the project area and its environment
7	Evaluation of all available information on the hydro-ecology of the area
8	Establishment of the cause of the damage if repair work is needed
9	Establishment of the objective and final appearance of the project
10	Selection of the live and dead vegetative material to be used
11	Selection of the construction method and type
12	Evaluation of the legal position (ownership, use, liability, etc.)
13	Evaluation of the hydro-engineering aspects of the project details
14	Final selection of the bioengineering technique to be implemented

*Orthophotos are aerial photographs corrected for distortion to conform to map accuracy at a given scale.

protection. Depending on the fluctuations of the water level, the river profile will be assessed and separated into zones of variable wetness. This, together with the results of the vegetation survey, will determine the vegetation type to be established in each zone of the planned embankment or watercourse bank.

It must be borne in mind that in nature there are rarely rigid and clear-cut boundaries between vegetation zones. Transitions are gradual; in some instances, a single vegetation type may cover all zones occurring in any one area. The approach to plant selection for the various zones therefore can not be too rigid.

Matching the various zones of the river corridor to the established vegetation zones will approximate the situation reflected in Fig. 1.1. A standard vegetation profile with regard to plant community distribution cannot be laid down, because the vegetation type may vary considerably within the region and the various reaches within the system. The final choice in what bioengineering technique to select for any given project can only be made within a framework of options.

The establishment of suitable and habitat adapted plants in the low water zone, such as pondweed (*Potamogeton*), Parrot's feather (*Myriophyllum aquaticum*), water crowfoot (*Ranunculus aquatilis*) and others, is still problematical, as the difficulties with regard to propagation, transport and planting of these sensitive semi-aquatic herbs have, as yet, not been overcome. Wave action and erosive forces are very pronounced in the middle bank, that is, the zone between the low and mean water level of the river. Depending upon local site conditions, various reeds may dominate the vegetation cover in this zone which is therefore often referred to as the 'reed zone'.

The shoots and leaves of the various reeds below water level disturb and slow the flow velocity, dissipate energy and cause sedimentation and aggradation; roots and rhizomes bind and stabilise the soil.

Reeds prefer sunny habits, a fact that must be taken into consideration when planning the selection of woody plants and their shading effect. The proper reed *Phragmites communis* is the best known of the reeds and also the most suitable for the protection of river and stream banks, and particularly lake shores. Reeds can withstand high flow velocities and are outstanding in their ability to withstand the wave action caused by passing powerboats because their dense interwoven roots and stolons are firmly anchored in the soil.

Depending upon the water level on site, the availability of suitable planting material, the time of the year and type of bank protection required, rootball, rhizome or shoot plantings may be employed (see Figs. 3.24 and 3.25; Plates 35–37); the planting of shoots and culms is the

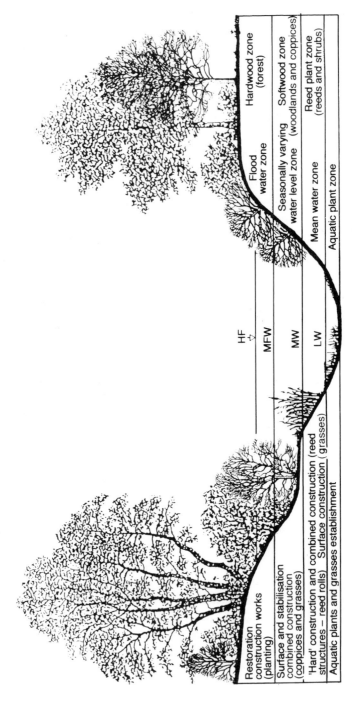

Key: HF, maximum flood water;   MFW, mean flood water;   MW, mean water level   LW, mean low water.

Fig. 1.1 Idealised watercourse cross-section showing water management and vegetation zones and appropriate bioengineering measures.

cheapest and fastest method, but has the disadvantage of being restricted to a four to six week long season during May and June. Underwater zones can only be planted successfully if the above water stems are not damaged during the full year following planting, because the roots obtain their oxygen requirements only from the vegetative parts above the water line. If the stems are damaged by mechanical action, the parts below the soil surface die back (Bittmann, 1953).

Reed canary grass (*Phalaris arundinacea*) is the only reed grass that does not cause rapid aggradation or sedimentation, but it has the ability to survive in polluted water. This makes it particularly suitable for the protection and stabilisation of the banks and shores of small water bodies. It spreads under water to a depth not exceeding 0.5 m, limiting its distribution to a relatively narrow margin. It is suitable for the banks of rivers with high flow velocities and widely fluctuating water levels and can be established by the use of rootballs, stolon cuttings or seed. The plants require a certain minimum flow velocity and come into their own, amongst other reeds, where, owing to large fluctuations of the water level, alternate dry and flooded conditions prevail. Planting material may be dug up in suitable areas of prolific growth, and transplanted throughout the year. It may be used in conjunction with quarry stone revetments.

Common clubrush or greater bulrush (*Schoenoplectus lacustris*) grows mainly in the aggradation zone of slow moving watercourses, preferring lake shores. It extends to a water depth of 0.4 m and grows actively throughout the whole year, even through winter. Therefore it should not be used in small water bodies, because its vigorous growth would restrict the profile. Of all the reeds and rushes, the clubrush is outstanding in its ability to oxygenate the water and to take up organic and inorganic compounds from the water. It takes up all traces of phenol and removes, per unit area, more pathogenic germs from the water than some drip-filter bodies (Seidel, 1965a; 1968; 1971).

Reed mace or bulrush (*Typha latifolia*), one of the tall reeds, is a pioneer in silted areas and can tolerate a water depth of up to 2 m. It is not suitable for use in small water bodies.

Some of the tall sedges are used in special sites with suitable habitats, as they are very sensitive to the depth of water. The success of plantings will hinge largely upon this factor, and the selection of the most suitable planting material should be made accordingly. There are plants with a narrow ecological amplitude, and others that will tolerate a wider range of conditions (Fig. 1.2).

The part of the upper bank above the mean summer flow level is very vulnerable to the erosion forces of wave action and ice drift. Grasses and

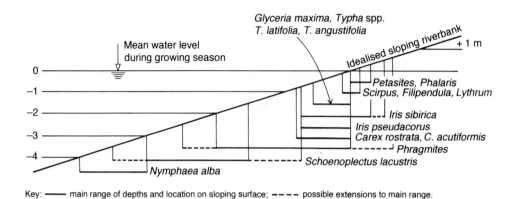

Key: ⎯⎯ main range of depths and location on sloping surface; ⎯ ⎯ ⎯ possible extensions to main range.

Fig. 1.2 Habitats for reeds and rushes versus depth of water.

shrubs, mainly willows, are the preferred vegetative protection for this zone.

The advantages of using grasses for the protection of riverbanks over other protection measures are numerous:

❑ Grass does not unduly retard or restrict the free passage of flood waters.
❑ A dense sward is relatively easily established, if the necessary conditions for rapid growth are met and the variety to be planted is suitable for the local conditions.
❑ Turf has a long lifespan.
❑ The establishment of a good grass cover is cheaper than other protection methods.
❑ Turf maintenance is very simple.
❑ The accumulation of floating debris is not a problem, and its removal easily achieved.

There are, however, several disadvantages:

❑ Maintenance is simple, but rather costly.
❑ The valley and watercourse are poorer for the absence of woody plants.
❑ The proliferation and growth of water plants is enhanced by the lack of shade.

In spite of these advantages, turf will maintain its importance as the dominant protection measure on the upper bank above the mean water level, on dykes and berms and similar flood control structures. The stability of a dense grass sward is extraordinarily great; it can withstand flow velocities of 1.8 m/s for prolonged periods, which may increase to 4.5 m/s for shorter periods. The permissible tractive force amounts to 105 N/m² (Linke, 1964).

Damage to established turf occurs and starts at weak spots such as cracks caused by drought, animal burrows, and cattle tracks and paths. In a closed sward, higher flow velocities cause the culms to bend, thus creating a zone of decreased velocities which actively protects the bed or bank from erosion even if the grass is relatively short.

Zones of watercourse banks subject to fluctuating water levels are preferably protected by shrub willows capable of throwing roots. The shrubs and trees in the zone above the mean flow level cause a decrease in the flow velocity close to the shoreline and the dense mass of roots reaching deeply into the subsoil bind the bank surface very effectively. The branches are meant to check the erosive force of the moving water and change part of the kinetic energy through turbulence to harmless friction, heat, oscillation and sound waves. Furthermore, the branches prevent the impact of floating ice on the shoreline. Trees and bushes that protrude above the top level of the watercourse bank reduce the flow velocity of flood waters during periods of inundation, thus protecting the endangered crown of the bank. In addition, debris and flotsam are retained by the bush belt, preventing their deposition in the areas behind the bank.

The branches and stems of all woody species planted in the zone of fluctuating water levels should be flexible because rigid and stiff trees and bushes interfere with the free flow of high flood waters, causing eddies and disturbed flow which often erode the watercourse bank, sometimes breaching them. All woody plants in this zone must therefore be radically pruned every five to eight years, taking care that through selective removal of branches and stems the closed bush formation is maintained.

Pre-conditions and limits for the use of shrub or bush willows include the following:

❑ The natural riverbed must be stable, because the willows cannot prevent bed scour or erosion occurring at the toe of the bank.
❑ The reach of the river to be protected must be within the zone where the planting material to be used occurs naturally.
❑ The slope of the riverbank should not exceed 1:3 and only in exceptional cases approach 1:2 or 2:3.
❑ Under natural conditions, willows will grow down to the level of the mean summer flow. Willow cuttings need at least one growing season

to develop a root system that can withstand flooding; willow plantings therefore should not be established below the mean summer flow level.
❑ Maximum tractive force levels should not exceed 100–140 N/m^2.

Willows are particularly well suited for vegetative stabilisation and protection methods:

❑ Of all woody species, willows will grow closest to the riverbed. They are capable of producing adventitious roots on several branches or long stems.
❑ All willow species may be rejuvenated by vigorous pruning to retain their bushy habit.
❑ Branches and long stems are very elastic and can withstand the action of ice drifts and rock slides.
❑ Very vigorous growth can withstand severe damage and rejuvenate very easily.

However, the following points need careful consideration in order to realise the full potential of willows:

❑ They are not very tolerant of shade. If planted in conjunction with other woody species, they soon lose their vigour. Maintenance operations on riparian willow plantations must aim at keeping competitive woody plants in check, particularly tree-size alders and spruce.
❑ The root system of willows is wide-spreading, but will penetrate to a great depth only in fairly permeable loose soil. They have a high oxygen demand, and do not tolerate a dense grass cover. For this reason it is not possible to reinforce an existing dense turf with willow cuttings to increase its erosion resistance. Regular pruning and cutting back will, however, increase root development.
❑ For optimum growth, willows need warm temperatures and plenty of moisture during the months of April and May. Above average rainfall and short duration flooding is most beneficial.
❑ Willows can tolerate total flooding for a period of eight days without deleterious effects. Partial flooding, that is, when several branches protrude above the water level, may last for several weeks without damage to the bushes.
❑ The total lifespan of willows is limited to about 40 years under natural conditions. If there is no competition from other woody plants, or if the bushes are radically pruned back on a regular basis, the lifespan may exceed 100 years.

❏ The various species are differentiated by their growth habit (see Fig. 2.1), habitat requirements (see Fig. 2.2) and suitability for use as live building material. In this regard, the right choice of planting material is essential if it is to take root readily and grow vigorously in order to fulfil the role with a minimum of care and maintenance (Schiechtl, 1992).

❏ The advantage of willows lies in their general robustness which can be maintained by suppressing the natural succession and keeping it permanently in check.

Watercourse bank protection measures based on use of vegetative material should be placed as low as possible down slope from the mean water level to avoid erosion starting at the toe which may lead to bank slumping in places. Protection measures that are confined to parts of the bank lying above mean water level will not be effective.

Chapter 2
Water Bioengineering Techniques

2.1 Definition

Bioengineering (Kruedener, 1951): an engineering technique that applies biological knowledge when constructing earth and water constructions and when dealing with unstable slopes and riverbanks. It is a characteristic of bioengineering that plants and plant materials are used so that they act as live building materials on their own or in combination with inert building materials in order to achieve durable stable structures. Bioengineering is not a substitute; it is to be seen as a necessary and sensible supplement to the purely technical engineering construction methods.

2.2 Function and effects

Even the most carefully planned civil engineering project cannot avoid bringing about certain changes in the landscape, thereby modifying the original terrain and causing temporary localised instability. The shaping and protection of such areas is of great importance and the latter can be achieved by vegetative means. In combination with technical methods, bioengineering techniques have the added advantage of effecting ecological, economic and aesthetic improvements. Depending upon the emphasis placed upon the construction type and method, various effects can be achieved (Table 2.1).

The adoption at a late stage of bioengineering techniques as remedial measures for the amelioration or rectification of any perceived shortcomings in project plan or flaws in the execution of civil engineering works is rarely, if ever, effective. Vegetative methods should best be integrated at the planning stage to supplement traditional civil engineering methods with the aim of improving their effectiveness. Under certain circumstances, bioengineering techniques may replace traditional methods entirely, if they are deemed to be more effective.

14

Table 2.1 Function and effects of water bioengineering techniques.

Geotechnical	• Protection of riverbanks and shores against erosion caused by flowing water and wave action. • Protection of watercourse banks against sheet erosion caused by rain, wind and frost. • Increased slope stability by the establishment of a soil–root matrix and modification of soil moisture content. • Protection against wind action and rock fall.
Ecological	• Modification of the extremes of temperature and moisture of the air in close proximity to the soil surface, thereby creating better conditions for growth. • Improvement in the soil–water relationships by drainage and water storage. • Increased soil and humus formation • Provision of habitats for animals and plants. • Shading of riverbanks and spawning areas by shrub and tree cover. • Water purification by retention of pollutants in the root zone. • Protection against wind action.
Economic	• Reduced costs of construction and maintenance. • Creation of usable areas for agriculture, silviculture and recreation.
Aesthetic	• Landscape harmonisation of alignment of major features. • Integration of structures into the landscape, rendering it more attractive. • Visual enhancement of the landscape by modifying structures.

The flow characteristic of any watercourse will be modified by the type, structure and extent of the vegetative cover of its banks. Improperly planned and poorly maintained trees and shrubs may interfere with the free flow of flood water. Dense shrub and tree growth on watercourse banks and riverbanks not only decreases flow velocities but also influences the distribution of velocities within the river profile (Felkel, 1960). The change in flow velocity distribution leads to a reduction in the discharge rate which is more pronounced in narrow riverbeds than in wide ones. Felkel has modified the formula by Manning–Strickler as applied to the mean discharge of river profiles with wooded banks by introducing the factor for retardation P_0/P (P_0 = wetted perimeter free of vegetation, P = wetted perimeter in metres). An example, using this factor, is given in Table 2.2 (DVWK, 1984). This has led to the recommendation that the banks of small rivers should only be planted to bushy willows if the minimum bed width exceeds 5 m, possibly limiting such plantings to only one bank (Swiss Federal Institute of Water Management, 1982).

The presence of woody vegetation on both riverbanks of natural and undisturbed river courses does not constitute an impediment to the

Discharge calculations:

$$Q_m = A_m \cdot V_m = A_m \cdot k_{GMS} \cdot \left(\frac{P_0}{P}\right)_m \cdot R_m^{2/3} \cdot S_b^{1/2}$$

1	2	3	4	5	6	7	8	9	10	11	12	13	14	15	16	17	18	19	20	21	22
Chainage in km	Transverse-section profile	Length of reach used in calculation	Bed level	Water surface level	Depth of water d	Difference: water surface levels	Transverse-sectional area of flow A	Average transverse-sectional area of flow (A_m)	Wetted perimeter P	Average perimeter P_m	Mean hydraulic radius R_m	Bed gradient S_b	Wetted perimeter P — Wooded bank	Wetted perimeter P — Bare bank	$\frac{P_0}{P}$	k_{GMS}	$\left(\frac{P_0}{P}\right)$ mean	$R^{2/3}$ mean	$S_b^{1/2}$	V mean	Q mean
km	–	m	m a.s.l.	m a.s.l.	m	m	m²	m²	m	m	m	–	m	m	–	m^{1/3}/s	–	m^{2/3}	–	m/s	m³/s
0 + 000	I	50	52.00	54.06	2.06	0.10	20.07	20.07	17.44	17.44	1.151	0.0020	5.56	11.88	0.681	30	0.681	1.098	0.0447	1.00	20.07
0 + 050	I	60	52.10	54.16	2.06	0.13	20.07	17.94	17.44	15.10	1.188	0.0022	5.56	11.88	0.681	30	0.709	1.122	0.0469	1.12	20.09
0 + 110	II	50	52.22	54.29	2.07	0.12	15.81	15.93	12.76	12.81	1.244	0.0024	3.35	9.41	0.737	30	0.738	1.157	0.0490	1.26	20.07
0 + 160	II		52.32	54.41	2.09		16.05		12.85				3.35	9.50	0.739	30					

Table 2.2 Example of discharge calculations for a watercourse with coppice banks.

discharge of flood waters, because there is enough space in the river corridor to cope with increased flow. In densely settled and regulated areas, river engineering based on vegetative means can be difficult due to lack of sufficient space.

2.3 Live construction materials

In water bioengineering the choice of live materials to be incorporated include:

❑ seeds of grasses, herbs and woody plants;
❑ parts of woody plants capable of vegetative propagation: cuttings, branches, whips and roots;
❑ parts of herbs and grasses capable of vegetative propagation: root divisions and rhizomes;
❑ rooted plants of grasses, herbs, shrubs and trees;
❑ topsoil sections complete with vegetative cover: grasses, herbs, small shrubs and very young trees;
❑ nursery produced grass and herbturf, some grid and mesh reinforced;
❑ tied bundles and mats of live plants and plant material: fascines, brush mats, reed rolls, wattle fences and mats.

The plant materials must meet certain standards, particularly with regard to origin, health, size and age.

2.3.1 Species selection

The proper selection of plant materials is based on four principles, which are to be carefully balanced or used in combination:

❑ objective of the measures;
❑ ecological make-up of the species;
❑ biotechnical properties of the species;
❑ origin (or provenance).

The objective of the measures taken is, in the first instance, the stabilisation of the bank or shoreline. At the same time, the bioengineering structures should be easy to maintain, and preferably lend themselves to economic use as pasture or woodland.

The ecological make-up or constitution comprises the effects with which plants react to their immediate environment. Plants and plant communities are therefore indicators of the conditions prevailing in their habitat. The plants best suited for use in bioengineering techniques have their origins in habitats similar to those where they are to be used. A detailed vegetation survey of the surrounding area will provide the best information on the location from where the plant material should be obtained. The use of such material will contribute substantially towards the success of the measures taken.

If it is necessary to remove vegetation from the construction site, turf and topsoil, together with larger plants, should be temporarily stored, and reinstated after completion of the works.

The selection of suitable plants is a prerequisite for the success of the measures taken; species with a wide ecological amplitude are particularly well suited to vegetative methods employed in bioengineering in water engineering. The use of unsuitable plant species may lead to failure. The following plant species are characterised by their wide ecological amplitude:

Trees	Grey alder (*Alnus incana*)
	European larch (*Larix decidua*)
	False acacia or black locust (*Robinia pseudacacia*)
	Goat willow or sallow (*Salix caprea*)
	Silver birch (*Betula pendula*)
	Common alder (*Alnus glutinosa*)
	Black poplar (*Populus nigra*)
	Scots pine (*Pinus sylvestris*)
Shrubs	Dogwood (*Cornus sanguinea*)
	Hoary willow (*Salix elaeagnos*)
	Fly honeysuckle (*Lonicera xylosteum*)
	Common osier (*Salix viminalis*)
	Privet (*Ligustrum vulgare*)
	Almond leaved willow (*Salix triandra*)
	Purple osier (*Salix purpurea*)
	Elder (*Sambucus nigra*)
	Black willow (*Salix nigricans*)
Grasses and legumes	Creeping bent (*Agrostis stolonifera*)
	Perennial rye grass (*Lolium perenne*)
	Bird's-foot trefoil (*Lotus corniculatus*)

Cocksfoot (*Dactylis glomerata*)
Red clover (*Trifolium pratense*)
Red fescue (*Festuca rubra*)
Sweet vernal grass (*Anthoxanthum odoratum*)
White clover (*Trifolium repens*)
Smooth meadow grass (*Poa pratensis*)
Kidney vetch (*Anthyllis vulneraria*)

The term 'biotechnical constitution' of a plant denotes the ability of plants or parts of plants to withstand the action of mechanical forces being exerted on stem or root. Plants with a high biotechnical rating should have the following properties:

❑ High resistance of stem and root towards mechanical forces. Of special importance is the tolerance for high flow velocities, large tractive forces and turbulent flow. Tolerance of temporary loss of soil cover alternating with sediment deposition and base load pressure (Table 2.3). Mechanical forces exerted by soil movement and the resulting tension and shear forces, snow pressure and rockfall should be equally tolerated by means of their root tensile strengths (Table 2.4).

Table 2.3 Maximum permissible mechanical forces in N/m^2 for structures immediately after completion and after three to four years' plant root development.

| | Force (N/m^2) | |
| | immediately | after 3–4 |
Construction material	after completion	seasons
Turf	10	100
Reed plantings	5	30
Reed roll	30	60
Wattle fence	10	50
Live fascine	60	80
Willow brush layer	20	140
Willow mat	50	300
Deciduous tree plantings	20	120
Branch layer	100	300
Coarse gravel and stone cover with live cuttings	50	250
Rip-rap with live branches	200	300
Rip-rap large quarry stone	—	250
Dry stone wall, stone pitching	—	600

Table 2.4 Tensile strength in N/mm^2 of plant roots less than 2 mm diameter.

Grasses	5–25
Herbs	3–60
Shrub-willows	10–70

❑ Tolerance of periodic flooding. Short-term flooding, lasting from a couple of hours to over two weeks, may occur every year several times in alluvial woodland, flood plains and forested wetland. The natural vegetation in such areas is adapted to these conditions. Vegetation from different habitats planted in periodically inundated areas would suffer badly or perish. Longer lasting flooding and stagnant water conditions are tolerated only by a few woody species, essentially by trees, e.g. white willow (*Salix alba*), crack willow (*S. fragilis*) together with one of its hybrids, *S. rubens*, and bay-leaved willow (*S. pentandra*). Sudden flooding is less well tolerated than a gradual rise of the water level. If the area is to be flooded artificially for whatever reason, the tree trunks should be surrounded with a layer of gravel up to the height of the expected water level to facilitate the formation of adventitious roots (Plate 1).

❑ Tolerance of sediment and gravel cover. Sediment cover caused by flood waters will kill most grasses and herbs, the more so if the deposited material is more than 100 mm thick and consists of clay and/or silt. Many woody plants will tolerate such conditions without any detrimental consequences. Willows and pines have survived gravel and rubble cover reaching 3 m, i.e. one-third of the total tree height, without showing signs of damage. Fine grained clayey material could, however, be less well tolerated.

❑ Ability to produce underwater roots. Several willows are able to produce dense floating masses of roots on the submerged part of the trunk. Such root masses dissipate the energy of the flowing water very efficiently, thus providing good protection from erosion for the bank.

❑ Soil binding properties which depend upon the type of root and density of the total root mass. Naturally, *in situ* germinated woody plants are capable of producing massive root systems which are usually widespread and, depending upon species, confined mainly to the topsoil layer (shallow rooted), or penetrating to greater depth (tap root systems). In a natural environment vegetatively propagated woody plants cannot strictly be separated into one or other of these types. Grasses with their very dense root masses are typical of the shallow rooted type. Herbs and some legumes form root systems that form a transition between the two types.

The most effective soil binding action takes place when the root system penetrates several soil layers; for best results therefore it is necessary to plant several different species.

❑ Ability to colonise subsoil and act as a pioneer plant in rubble and soil-substitute materials, preparing the site for other plant communities.

❑ Ability to improve the soil in the absence of further human interference, thus preparing the site for the smooth transition for higher plant associations. Improved soil cohesion and fertility and the creation of a suitable soil micro-climate will bring about this change. Vigorous growth and increased biomass production are important factors. Very important from the practical point of view are plants that are able to enrich the soil with nitrogen, either with the aid of root nodules or the leaves shed in autumn. Legumes and alders are able to fulfil this function.

❑ Tolerance of high salt concentrations. This is important in coastal regions and where rock salt is spread in relatively large quantities to improve traffic safety during winter.

2.3.2 Vegetation zones and plant origin

Meusel (1965) and Tschermak (1961) have grouped the potential natural vegetation of any given region into regional vegetation zones or vegetation areas, respectively. In the present context, the broader divisions comprise the Central Alpine region, the main Continental region, the Atlantic (Oceanic) influenced northern, and the Mediterranean influenced southern fringe of the Alps. Within these floral zones, natural plant distribution and growth vigour is basically determined by the elevation of the area above sea level. The selection of plant materials for construction purposes must therefore take due cognisance of its provenance and the elevation of its source location.

In principle, the aim should be to use only plants and plant materials from areas in close proximity to the construction site. The higher the elevation of the site, the more important is the observation of this basic rule.

2.3.3 Plant propagation

Following the important principle of plant provenance and the necessity to have large quantities of plant materials on hand, economic considerations demand the easy propagation of plants involved to prevent bottlenecks and shortages during construction. The propagation of biological building materials is either generative using seeds or vegetative using plant parts capable of producing shoots and/or roots. It could be of great advantage to the engineer to have recourse to commercially produced grass seed mixes which are made up of those species that constitute the natural plant succession for any given site. Seed collection from indigenous plants growing in the wild all but ceased a long time ago as it is not economically viable, but it is still used as a last resort for special projects.

Hayseed, collected from hay ricks and barns, is of importance locally and much sought after. Site conditions are, however, not the sole criteria for the choice of the seed mix; the cost factor needs careful consideration.

The practice of obtaining seeds of grasses, herbs and woody plants from local commercial sources has been found to be satisfactory, as long as they meet the requirements of hardiness and expected lifespan. Most of the grass seed varieties offered by the trade were bred for agricultural purposes of ornamental lawns, and in many cases are not suitable for use by the engineer in the protection and erosion control of waterway earth structures. It should be borne in mind that seed mixes should be made up of as many species as possible, because such mixes correspond more to the prevailing natural conditions, and provide a more stable vegetative cover. The use of seed of a single variety is only moderately successful.

Several countries recommend for certain regions, or even isolated local habitats, so-called standard seed mixes, and seed merchants offer, to the user, such mixtures blended according to their experience. It is preferable that seed mixes are prepared on the advice of an expert. Commercially available or recommended seed should be tested with regard to its suitability for the site conditions and correct species composition. The relatively small cost of this is worthwhile.

The preparation of seed mixes of woody plants needs special attention and experience. Apart from the woody species, the mix must contain seeds of grasses and herbs because the seeding does not aim at the establishment of some sort of shrubbery or woodland, but primarily at the stabilisation of the earth structure slopes, e.g. banks of watercourses. The seeds of grasses and herbs that would compete to the detriment of the woody species obviously must be excluded from the mix.

Table 2.5 at the end of this chapter lists most of the seed varieties suitable for use in water bioengineering construction together with their properties. Woody plants capable of vegetative propagation constitute the most important material for ground stabilisation and combined construction techniques (Sections 3.2 and 3.3). Such plant material is used in various sizes:

❏ Cuttings and stakes or truncheons are unbranched stems and shoots 250–600 mm long and approximately 20–80 mm diameter.
❏ Branches and twigs are a minimum 600 mm long and of varying thickness.
❏ Switches are slender and flexible minimally (or scarcely) branched shoots of 1–1.5 m long.
❏ Poles are straight poorly branched stems 1–2.5 m long.

For best results, the live wood should be fairly thick and long because with an increase of plant mass, bud formation, sprouting and rooting increase proportionally; experience has shown that 20–80 mm diameter branches at the cut end are best. Thin branches and switches are subject to rapid desiccation and are usually used in conjunction with more substantial parts. The required plant material may be obtained from the following sources:

❏ similar nearby habitats with similar ecological conditions;
❏ stems and branches from previously established bioengineered shrub and tree cover, obtained by maintenance pruning, thinning, coppicing and pollarding;
❏ as a last resort, from forest nurseries and similar commercial outlets, if natural scrub or woodland is not available.

The best time for the cutting of suitable material is during the dormancy period, i.e. the time between leaf shed and bud burst (October to April). Shrubs and young trees are cut just above ground level, older trees have their branches lopped off. To produce a clean cut, saws or secateurs are used.

All branches are transported whole to the construction site to prevent them from drying out, and cut to the required length and shape only on site, for immediate use. Any delay should be avoided and care must be taken to establish a good soil cover to prevent desiccation.

If there is any time lapse between cutting and use on site, plant material cut during the dormancy period may be stored provided it is protected from drying and bacterial heating. This can be achieved by storage in snow or cold rooms (0–1°C), submergence in flowing water of a temperature below 15°C, or cover in special PVC bags and foils. Antidesiccant chemical may be used fairly successfully. If the buds show signs of swelling, further storage is impossible.

In water engineering it is possible to use plants and plant material capable of vegetative propagation which was cut during the growing period, provided it is planted immediately after collection in moist or permanently wet soil.

There are approximately thirty woody species in Central Europe that are capable of vegetative propagation, predominantly willow species (Figs. 2.1 and 2.2). Of these, ten are of specialist use in the montane and sub-alpine regions of the Alps (Schiechtl, 1992). In Table 2.6 at the end of this chapter all thirty species are described and their properties listed.

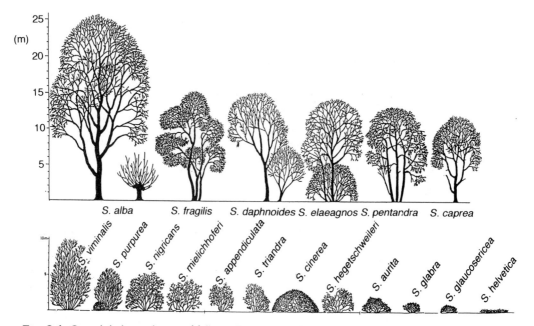

Fig. 2.1 Growth habit and sizes of full-size European indigenous tree and shrub willows.

All the species shown in Table 2.6 may be employed as rooted saplings grown from seed. Rooted young plants of non-vegetatively propagating trees and shrubs are used in the construction of hedge layers (see Section 3.2.3) or as a supplement in the hedge-brush layer method (see Section 3.2.3.3). For this purpose pioneer plants, which can tolerate sediment cover and produce roots of high tensile strength, are preferred; conifers are not suitable.

The most suitable shrub and tree species and some of their more important properties are listed in Table 2.7 at the end of this chapter.

Most woody plants occurring in any given floral area are suitable for

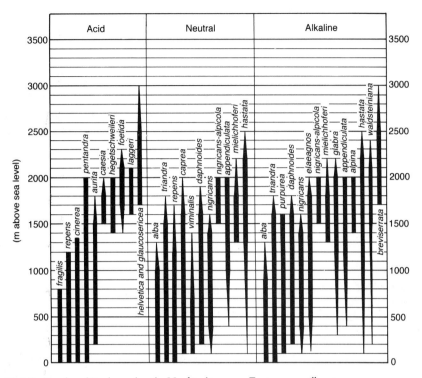

Fig. 2.2 Distribution by altitude and soil pH of indigenous European willows.

horticultural or landscaping purposes, but only a few are suitable for use in civil engineering structures. These are essentially the vegetatively propagated species mentioned in Table 2.6 and the species listed in Table 2.7. Plants that are not suitable for use in bioengineering construction may, however, be planted at a later stage after the immediate objective of protecting and stabilising the damaged or endangered site has been achieved.

2.4 Preliminary work

Preliminary work or site preparation comprises measures that are needed to make the site safe and to protect the workforce. These are essentially temporary measures such as diversions, scaffolding, barriers and fences. Areas where loss of soil and vegetation has occurred due to erosion need corrective reshaping before the actual protection measures can be implemented, abrupt surface features such as ridges, ribs and hollows must be graded flat and hollowed out, and crest overhangs on under-washed riverbanks must be rounded off, thus re-establishing acceptable slope gradients to prevent further erosion. Abrupt changes in surface shape are sites of constant erosion and sources of sediment and therefore should always be rounded off to eliminate problems in the future. This work can be done by water-jetting or mechanically.

The preservation or salvaging of reusable materials, dead and live vegetation and other locally occurring useful materials forms a very important aspect of bioengineering construction. Prior to commencement of work, arrangements must be made to ensure that this can be done. The more extreme the conditions on site, the more vital this matter is. After the partial or full completion of the bioengineering works, previously removed or translocated rooted vegetation layer should be replaced within the construction area, augmented if necessary by additional bioengineering measures.

2.5 Selection of bioengineering construction approach and technique

The selection of the available plant material will influence the decision as to which construction approach and technique to use (see Section 2.3.1). In addition, the time of the year must be considered to comply with the appropriate optimum planting date. The following summarises the factors that decide the choice of the construction approach and technique adopted:

- ❑ Objective of the scheme: the immediate short-term aim is the stabilisation and protection of the shoreline, or the watercourse bed and the riverbed. Further goals are the creation of necessary conditions for the establishment of aquatic life and land animals, planting of easily maintained riparian woodland and forest or reed belts and/or turf. The impact of such works on the landscape must be considered also.
- ❑ Anticipated technical effectiveness: are vegetative means sufficient on their own or is there a need for combined construction techniques?
- ❑ Availability of live materials: which of the plants used in the measures can be obtained from nearby local habitats? Which have to be brought from further afield, but are suitable for establishment of the pioneer vegetation cover?
- ❑ Time of year: construction methods that require the use of plants capable of vegetative propagation are confined to the dormancy period (late autumn and winter, for exceptions see Section 2.3.3).

2.5.1 Construction timing (Fig. 2.3)

Construction timing is largely dependent upon the growing cycle of the vegetative material; if vegetatively propagated plants are to be used, the construction period is restricted to the dormancy period (October/November to March/April). For exceptions (sourcing and use during the growing period) see Section 2.3.3. Grasses are sown any time during the growing period; woody species during spring or in autumn for best results. Container or pot plants may be planted throughout the year.

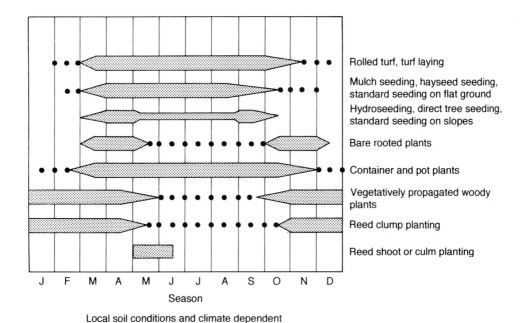

Rolled turf, turf laying

Mulch seeding, hayseed seeding, standard seeding on flat ground

Hydroseeding, direct tree seeding, standard seeding on slopes

Bare rooted plants

Container and pot plants

Vegetatively propagated woody plants

Reed clump planting

Reed shoot or culm planting

J F M A M J J A S O N D

Season

Local soil conditions and climate dependent

Fig. 2.3 Installation schedule for bioengineering techniques, valid for Western and Central Europe, including Alpine regions, and other temperate zones of the Northern hemisphere.

2.5.2 Limits of application

The use of vegetative materials has biological, technical and time constraints.

❑ *Biological constraints:* areas that are unsuitable for certain plants; limits of distribution; excessive pollution.
❑ *Technical constraints:* slope stabilisation is only feasible if the substratum can support root growth. Deeper seated ground movement below the surface level can be indirectly influenced by bioengineered measures by decreasing the amount of soil water through increased evapotranspiration from a dense vegetative cover and root reinfor-

cement of soil. Further limits are excessive flow velocities, tractive forces and turbulence.

❑ *Time limits:* implementation during either the dormancy or the growing period.

These limitations highlight why bioengineering techniques in civil engineering are not a substitute for, but a supplement to, conventional purely technical methods.

2.6 Construction costs

The costs mentioned in conjunction with the various construction methods are those of projects carried out to date. In order to eliminate the fluctuating value of money, the construction costs are expressed in work hours. For direct cost comparisons of the various construction methods see Fig. 2.4. Costs will vary with site conditions, availability of construction materials and company structure. Accurate costings are therefore only possible when special tender conditions are available.

The low cost of bioengineering methods compared with those of classical civil engineering methods is often cited as the essential advantage of the former over the latter. This is not always the case, and occasionally bioengineering methods may be more expensive.

Much work has been done in many countries to ascertain the cost of bioengineering techniques expressed as a proportion of the total construction costs. It is of interest to note that these costs vary considerably, depending upon the amount and nature of the conventional construction work involved. As a rule, the proportional cost for the vegetative part of the project will increase with the desired degree of nature preservation and natural site appearance to be achieved, but, at the same time, the total project costs will decrease.

The maintenance costs are very much lower than the costs for the rehabilitation of mechanical or inert measures.

It is imperative that care and maintenance operations with all bioengineering techniques are meticulously carried out during the first few years after completion to ensure the success of the project. It is of advantage to make provision for and include this maintenance period in the original tender specifications. When the vegetation has established itself and becomes fully functional, nature will take over and all further input is confined to periodic maintenance work to ensure stability and efficiency (see Chapter 6). If the live construction material selected meets the ecological requirements of the site, maintenance will only be required at long intervals, and costs therefore kept to a minimum.

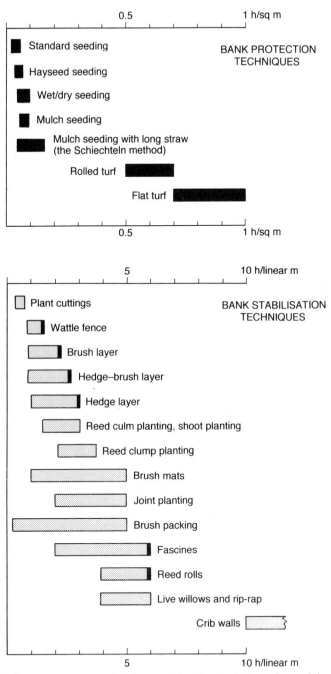

Fig. 2.4 Cost comparisons of various construction techniques (in work hours).

Table 2.5 Seeds of grasses, herbs and woody plants suitable for water bioengineering construction. (Source: Ehrendorfer, 1973.)

Latin and English name	Plant association*, habitat and properties	Morphology: stem and root	1000 grain wt	Seeds/g
Grasses				
Agropyron repens (Couch grass)	Agropyro–Rumicion, Artemisietea, Chenopodietea, Festuco–Brometea, var. glaucum in Mesobromion and Erico–Pinion. Arable weed, not to be used near arable land, needs light. Dry to moist, fertile; up to 900 m a.s.l., perennial.	Stem 0.2–1.5 m Root up to 0.8 m rhizomatous	4	260
Agrostis canina (Bent grass)	Carici–Agrostidetum. Fens and bogs, peat areas, wet woodland, pioneer plant of open wet soil. Subsp. canina on moist, subsp. montana on dry habitat of poor fertility; up to 1100 m a.s.l., perennial.	Stem 0.2–0.6 m Root 0.2 m	0.05	20000
Agrostis gigantea (Black bent, red top)	Calthion, Glycerion, Caricetum fuscae, Molinio–Arrhenatheretea, Phragmitetalia. Riverbanks and lakeshore, wet woodland; up to 1400 m a.s.l., unsuitable for lawns, biennial to perennial.	Stem 0.4–1 m Root 0.3 m with short rhizomes	0.05	20000
Agrostis stolonifera (Creeping bent)	Agropyro–Rumicion, Plantaginetalia. Riverbanks, ditches, wet areas, moisture indicator, pioneer plant, dense sward, can stand heavy grazing, best grass for golf links in the northern, moist and cool areas; up to 1800 m a.s.l., perennial.	Stem 0.1–0.7 m stoloniferous, rooting at the nodes Root 0.3 m	0.05–0.09	11000–20000
Agrostis tenuis (Common bent, brown top)	Festuco–Cynosuretum, Polygono–Trisetion, Arrhenatheretalia, Nardo–Callunetea, Sedo–Scleranthetea. Grassy patches in woodland, moist grassland in mountainous regions, moors, recently cut forest areas, indicates acid and very poor soil conditions; valuable grass for mountainous terrain, does well in grass mixtures; up to 2200 m a.s.l., perennial.	Stem 0.2–0.4 m Root up to 0.5 m	0.06	18000

a.s.l. = above sea level.
*These plant associations are specific to Alpine regions.

Species	Description	Dimensions		
Alopecurus pratensis (Foxtail)	*Arrhenateretalia, Calthion, Filipendulo–Petasition, Molinio–Arrhenatheretea.* Riverbanks, alluvial areas, resistant to late frosts, overwatering, tolerant of long lasting snow cover, needs fertiliser and irrigation, then suitable for poor soils; up to 1800 m a.s.l., moisture indicator, does not stand up to heavy grazing, perennial, loosely tufted.	Stem 0.3–1 m Root up to 0.2 m	0.7–0.9	1100–1400
Anthoxantum odoratum (Sweet vernal grass)	*Molinio–Arrhenatheretea, Nardo–Calunetea, Cariceta curvulae, Quercion roboris.* Meadows and pastures, open woodland, poor mountain meadows together with *Festuca rubra, Agrostis tenuis*; up to 2500 m a.s.l., grows during winter, can stand irrigation, indicates poor soil, short lived, perennial.	Stem 0.3–0.5 m Root up to 0.5 m	0.6	1700
Arrhenatherum elatius (Oat grass)	*Arrhenatherion elatioris, Calamagrostidion arundinacii.* Moist and fertile meadows, at higher elevations replaced by *Trisetum sp.*, main species of fertilised meadows, avoids wet heavy and compacted soils, sensitive to moist–cool conditions; up to 1500 m a.s.l., perennial, loosely tufted.	Stem 0.5–1.5 m Roots tough, deep rooting	3.3	300
Avena sativa (Oats)	Best cover crop in humid habitats, frost tender, high water demand, desiccates the soil; up to 1600 m a.s.l., annual, at high elevations in the mountains, bi-annual to perennial.	Stem up to 0.8 m Root up to 0.3 m	33.3	30
Avenulla flexuosa (Oat grass, flexible)	*Vaccinio–Piceion, Eu–Vaccinio–Piceion, Epilobion angustifolii, Luzulo–Fagion, Quercion roboris, Nardo–Callunetum.* Acid humus pioneer on poor soils, shade tolerant; up to 2200 m a.s.l., perennial.	Stem 0.3 m, slender Root up to 0.5 m, tough	1.25	1600

Latin and English name	Plant association*, habitat and properties	Morphology: stem and root	1000 grain wt	Seeds/g
Grasses *continued*				
Brachypodium pinnatum (Chalk false brome)	*Meosbromion, Cirsio–Brachypodion, Festuco–Brometea, Nardetalia, Molinion* dry, *Erico–Pinion, Cephalanthero–Fagion.* Indicates base-rich soil and deteriorating woodland conditions, enhanced by burning, weakened by fertiliser application; up to 1600 m a.s.l., perennial.	Stem 0.6–1.2 m Root up to 0.5 m, tough	4.6	220
Bromus erectus (Upright brome)	*Xerobromion, Mesobromion, Molinion, Arrhenatherion, Salvia–Trisetetum.* Semi-arid grassland on limestone, in the south on gneiss and serpentine, pioneer, intolerant of fertiliser and irrigation, avoids shade and wetness, but very resistant to dry heat; up to 1400 m a.s.l., perennial, tufted.	Stem up to 0.6 m Root up to 0.8 m, tough	4.6	220
Bromus inermis (Hungarian brome)	*Cirsio–Brachypodietum, Arrhenatherion, Sisymbrion.* Pioneer, very drought and cold resistant; up to 1200 m a.s.l., perennial.	Stem 0.3–1.4 m Rhizomatous, deep rooted	2.2–3.3	300–450
Bromus mollis (Soft brome)	*Bromo–Hordeetum, Sisymbrion, Arrhenatherion.* Grassland weed, indicates poor soil, suitable cover crop on dry sites; up to 1000 m a.s.l., annual.	Stem 0.2–0.8 m Root up to 0.2 m	3	325
Cynodon dactylon (Couch grass)	*Polygonion avicularis, Cynosurion.* Pioneer for the quick stabilisation of sandy soil in low altitudes, starts late in the season, frost hardy, pasture grass, turns brown in winter; up to 1000 m a.s.l., perennial.	Stem procumbed with long stolons Rhizomatous	0.3–0.58	1700–3500
Cynosurus cristatus (Crested dog's tail)	*Cynosurion, Arrhenatherion.* Fertilised permanent pasture, indicates heavy soil, frost tender, shade tolerant for pastures and meadows; up to 1500 m a.s.l., perennial.	Stem 0.3–0.6 m	0.5–0.58	1700–2000

Species	Description	Stem/root		Altitude
Dactylis glomerata (Cocksfoot)	*Arrhenatherion, Mesobromion*. All round pioneer, vigorous when fertilised; up to 1900m a.s.l., perennial.	Stem 0.5–1m Tufted, root up to 0.4m	1.2	900
Deschampsia caespitosa (Tufted hair grass)	*Molinion, Filipendary–Petasition, Calthion, Fagion, Alno–Padion*. Intermittent moist locations in woodland and meadows, wet places and spring horizons, marshes; up to 2800m a.s.l., perennial, vigorous, stiffly tufted.	Stem 0.3–0.8m Root up to 1m	0.25	4000
Festuca arundinacea (Tall fescue)	*Potentillo–Festucetum arundinaceae, Agropyro–Rumicion, Molinion, Calthion*. Indicates soil compaction and wetness, tufted, tolerant of trampling, suitable for pathways and terraces in vineyards and orchards; perennial.	Stem 0.6–1.5m Rhizomatous, deep rooted	1.62–1.9	530–620
Festuca ovina (Sheep's fescue)	*Festuca–Brometea, Sedo–Scleranthetea, Molinio–Arrhenatheretea, Quercion roboris, Pinion sylvestris*. Dry, poor grassland on soils derived from acid rocks, indicates degradation in forests; up to 2300m a.s.l., perennial.	Dense, short grass Stem 0.15–0.4m Root up to 0.5m	0.5	2000
Festuca pratensis (Meadow fescue)	*Molinio–Arrhenatheretea, Mesobromion*. Fertile meadows and pastures, prefers heavy and moist soils, winter hardy, sensitive to over-use; up to 1600m a.s.l., perennial.	Stem 0.3–1.2m Loose tufts, shallow rooted	2	500
Festuca rubra subsp. rubra (Red fescue)	*Molinio–Arrhenatheretea*. Montane meadows and pastures, coniferous wood and deciduous forests, drought and wetness sensitive; up to 2000m a.s.l., perennial.	Stem 0.2–0.7m Runners, root up to 0.5m	0.9–1	1000–1100
Festuca rubra subsp. commutata (Red fescue)	*Nardetalia, Cynosurion, Polygono–Trisetion*. On acid soils, replaced by *Nardus stricta* if over used; up to 2000m a.s.l., perennial.	Stem 0.1–0.6m Dense sward, root up to 0.5m	1	1000
Festuca tenuifolia syn. *capillata* (Fine-leaved sheep's fescue)	*Nardo–Galion, Thero–Airion, Quercion robori–petraeae, Castaneta*. On acid sandy soils, indicates deterioration in forests; up to 1000m a.s.l., perennial.	Stem 0.2–0.3m Slender, shallow rooted	0.4–0.6	1700–2500

Latin and English name	Plant association*, habitat and properties	Morphology: stem and root	1000 grain wt	Seeds/g
Grasses *continued*				
Festuca trachyphylla syn. *longifolia* (Rough-leaved fescue)	*Festuco–Sedetalia, Xerobromion, Seslerio–Festucion, Koelerion glaucae.* Origin in the low lying areas of northern Germany on sandy soils, widespread in Central Europe and England; up to 1000 m a.s.l., perennial.	Stem 0.1–0.15 m Very shallow rooted	0.6	1700
Holcus lanatus (Yorkshire fog)	*Arrhenateretalia, Molinion, Calthion.* Indicates acid soils low in nitrogen, frost tender, green during winter, in years of good rainfall very prolific on poor soils; up to 900 m a.s.l., perennial.	Stem 0.3–1.6 m Tufted, root up to 0.4 m	0.3–0.4	2500–3500
Holcus mollis (Creeping soft grass)	*Quercion robori-petraeae, Nardo–Callunetea.* Arable lands and ploughed out pasture, wet soil areas, never on calcareous soil, sandy soils, troublesome weed in gardens and arable land, starts growing later than *Holcus lanatus*; up to 1500 m a.s.l., perennial.	Stem 0.3–1.6 m Long rhizomes, root up to 0.4 m	0.25	4000
Lolium multiflorum subsp. *italicum* (Italian ryegrass)	*Bromo–Hordeetum, Sisymbrion, Arrhenatherion.* Only in Atlantic areas under moist mild climatic conditions; frost damage below –5°C, needs potash-rich soils, vigorous after cuts, unsuitable for permanent meadows; up to 1700 m a.s.l., annual to bi-annual.	Stem 0.3–0.9 m Root up to 0.8 m	2.2	470
Lolium perenne (Perennial ryegrass)	*Lolio–Cynosuretum, Polygonion avicularis.* Tolerant to repeated cutting and trampling, pioneer of well-aerated and moist soil, needs fertiliser; up to 1000 m a.s.l., but in certain locations up to 2300 m, fast growing perennial.	Stem 0.3–0.7 m Dense tufts, root up to 1.2 m	1.4–2.2	470–720

Species	Ecological description	Stem/roots		
Phleum pratense (Timothy)	*Cynosurion, Arrhenatheretalia.* Pastures, tolerates cold climate, wetness and prolonged snow cover, wind; grazing increases vigour and yield; up to 2600m a.s.l., perennial.	Stem 0.2–1m Loose tufts, roots delicate with short runners	0.5	2000
Poa annua (Annual meadow grass)	*Plantaginetalia majoris, Polygonion avicularis, Cynosurion, Chenopodietea, Secalinitea.* Tolerates trampling and heavy fertiliser applications; up to 3000m a.s.l., annual to perennial.	Stem 0.02–0.35m Dense low sward, shallow roots	0.5	2000
Poa compressa (Flat-stalked meadow grass)	*Alysso–Sedion, Tussilaginetum, Cirsio–Brachypodion, Festuco–Sedetalion.* Not on soils derived from acid rocks; up to 1800m a.s.l., green over winter, perennial.	Stem 0.2–0.4m Loose tufts with runners, root up to 0.2m	0.15–0.18	5500–6500
Poa nemoralis (Wood meadow grass)	*Carpinion, Fagion, Quercion pubescentis-petraeae, Prunetalia.* On heavy soils, grows early after melting snow, very shade tolerant, not to be planted in pure stands, does not form close sward; up to 2300m a.s.l., perennial.	Stem 0.2–0.9m Tufted, shallow rooting	0.6	1700
Poa palustris (Swamp meadow grass)	*Phragmition, Magnocaricion, Phalaridetum, Calthion, Alnion.* On riverbanks, early, resistant to late frost; up to 1500m a.s.l., perennial.	Stem 0.3–1.2m Shallow rooting	0.2–0.25	4000–5200
Poa pratensis (Meadow grass)	*Molinio–Arrhenatheretea, Mesobromion, Festuco–Brometea.* Important constituent of meadows and pastures, wide ecological amplitude, hardy, long living, grows early in spring, very suitable for first seedings; up to 2300m a.s.l., perennial.	Stem 0.15–0.9m Dense tufts, rhizomatous roots to 0.65m	0.22–0.31	3200–4500
Poa trivialis (Rough meadow grass)	*Calthion, Filipendulo–Petasition,* moist *Arrhenathereta.* On spring horizons, fens, sensitive to dry air and soil, frost and prolonged snow cover, resistant to heavy grazing, responds well to organic fertilisers; up to 1600m a.s.l., long life, forms dense sward.	Stem 0.3–0.9m Surface runners, shallow rooted	0.18–0.33	3000–5500

Latin and English name	Plant association*, habitat and properties	Morphology: stem and root	1000 grain wt	Seeds/g
Grasses *continued*				
Puccinellia distans (Sea meadow grass)	*Juncetalia Blysmo–Juncetum, Puccinellion maritimae.* Saline soils, manure heaps, cattle pens, solonetzic soil; perennial.	Stem 0.15–0.5 m Tufted, root up to 0.25 m	0.24	4200
Trisetum flavescens (Yellow oat)	*Polygono–Trisetion, Arrhenatherion.* Fertile meadows in montane and subalpine regions, frost sensitive, resistant to repeated cutting; up to 2300 m a.s.l., perennial.	Stem 0.3–0.8 m Loosely tufted, root up to 0.4 m	0.26–0.5	2000–3800
Herbs and legumes				
Achillea millefolium (Yarrow, milfoil)	*Arrhenatheretalia, Nardetalia, Mesobromion.* Meadows and pastures, drought resistant, indicator of fertile soil, medicinal use; up to 1900 m a.s.l., perennial.	Stem 0.2–0.6 m Rhizomatous, root up to 4 m	0.15	6700
Chrysanthemum leucanthemum (Ox-eye, dog daisy)	*Arrhenatheretalia, Molinietalia, Mesobromion.* Pioneer on loose, well-aerated, immature soils, indicates poor fertility in meadows; up to 2200 m a.s.l., perennial.	Stem 0.3–0.6 m Root up to 0.6 m	0.3–0.38	2600–3300
Pimpinella saxifraga (Burnet saxifrage)	*Xero–Mesobrometea, Nardo–Galion, Erico–Pinion, Festuco–Brometea.* Indicates poor, dry soil, useful in sheep pasture on calcareous soil; up to 2300 m a.s.l., perennial.	Stem 0.15–0.5 m Root up to 1.3 m deep, 8–10 m long	1.5–9	110–670
Plantago lanceolata (Plantain)	*Molinio–Arrhenatheretea.* Worldwide distribution on many soil types, even in waterfowl runs, very sensitive to herbicides; up to 1800 m a.s.l., perennial.	Stem 0.05–0.5 m Root up to 0.6 m	1.65	625

Species	Description	Size		
Sanguisorba minor (Salad burnet)	*Mesobromion, Festuca–Brometea, Arrhenatherion, Erico–Pinion, Brometalia erecti.* Pioneer plant, root fungus symbiosis, medicinal use, spice; up to 1200m a.s.l., perennial.	Stem 0.3–0.6m Root up to 1.5m	1.3–9	110–815
Anthyllis vulneraria (Kidney vetch)	*Mesobromion, Cirsio–Brachypodion, Xerobromion, Molinion, Erico–Pinion, Arrhenatherion.* Cannot take fertiliser application or irrigation, pioneer, frost and drought resistant; up to 2000m a.s.l., perennial.	Stem 0.1–0.5m Root over 1m	2.5	400
Coronilla varia (Crown vetch)	*Mesobromion, Arrhenatherion, Onopordion.* On dry sunny slopes; up to 900m a.s.l., perennial.	Stem 0.3–1.2m Root up to 0.9m	4	260
Lotus corniculatus (Bird's foot trefoil)	*Mesobromion, Trifolion medii, Arrhenatheretalia, Molinion.* Semi-dry turf, fertile meadows and pastures, prefers calcareous soils, high temperature resistance; up to 2300m a.s.l., perennial, persists for 20 years.	Stem 0.05–0.6m Tap root up to 1m	1–1.3	750–1000
Lotus uliginosus (Large bird's foot trefoil)	*Calthion, Molinion,* moist *Arrhenatheretea, Alno–Padion.* Moist meadows and pastures, spring horizons, bog plant, indicates nitrogen-rich soil; up to 1000m a.s.l., perennial.	Stem 0.3–0.9m Root more than 1m	1–1.3	750–1000
Lupinus albus (White lupin)	Up to 600m a.s.l., annual.	Stem 0.2–1m Root up to 0.75m	33.3	30
Lupinus luteus (Sweet lupin)	Up to 1400m a.s.l., annual.	Stem 0.3–1.2m	33.3	30
Lupinus polyphyllus (Garden lupin)	*Sambuco–Salicion.* On woodland fringes and clearings; up to 1400m a.s.l., perennial.	Stem 1–1.5m Root more than 1m	22.2	45
Medicago falcata (Sickle medick)	*Festuco–Brometea, Geranion sanguineae–Brachypodion, Arrhenatherion.* Not cultivated as it becomes woody. *Medicago varia* (=*M. falcata* × *sativa*) traded under the name 'lucerne'; up to 1100m a.s.l., perennial.	Stem 0.2–1m Deep rooted	2	500

Herbs and legumes continued

Latin and English name	Plant association*, habitat and properties	Morphology: stem and root	1000 grain wt	Seeds/g
Medicago lupulina (Black medick)	*Mesobromion, Caucalion, Lolio–Cynosuretum, Arrhenatheretalia.* Dry meadows of good fertility, indicator of dry habitat, undemanding pioneer, prefers calcareous soil, frost resistant, can be heavily grazed; up to 1500 m a.s.l., annual to perennial.	Stem 0.1–0.6 m Thin tap root, up to 0.5 m	1.8–2.3	435–550
Medicago sativa (Lucerne)	*Mesobromion* and dry *Arrhenatherion.* Native of Iran, today only available as hybrid, sensitive to late frosts; up to 1000 m a.s.l., perennial.	Stem 0.3–1.2 m Very tough tap root, 2.5–5 m (up to 10 m)	0.7–2.5	400–600
Melilotus albus (White melilot)	*Echio–Melilotetum.* Drought resistant, becomes woody, needs mowing; up to 1800 m a.s.l., bi-annual.	Stem 0.3–1 m Thick tap root, 0.7 m	1.8	570
Melilotus officinalis (Common melilot)	*Echio–Melilotetum, Tussilaginetum, Caucalion.* Used medicinally; up to 1000 m a.s.l., bi-annual.	Stem 0.3–1 m Tap root, up to 0.75 m	1.8	570
Onobrychis viciifolia (Common sainfoin)	*Mesobromion, Brometalia erecti.* Dry soil indicator in *Arrhenatheretum*; important feed plant on dry, clayey calcareous soil, sensitive to grazing pressure; up to 2000 m a.s.l., lasts 4–6 years.	Stem 0.1–0.7 m Root 1–4 m	20–29	35–50
Phacelia tanacetifolia	Up to 1000 m a.s.l., useful as green crop, annual.	Stem up to 0.7 m Root up to 0.2 m		
Pisum sativum (Garden pea)	Up to 1000 m a.s.l., cover and green crop, annual.	Stem 0.5–2 m Tapering tap root	143–500	2–7
Trifolium dubium (Lesser clover)	*Arrhenatheretea, Cynosurion, Arrhenatheretum.* Fertile meadows and pastures, needs heavy nitrogen dressings; up to 1000 m a.s.l., annual to bi-annual.	Stem 0.05–0.35 m Root up to 0.2 m	0.5–0.55	1850–2000

Species	Habitat/notes	Dimensions		
Trifolium hybridum (Alsike clover)	*Arrhenatherion, Bromion racemosi, Molinion, Calthion, Agropyro–Rumicion.* Pioneer on moraines, tolerates moist and cool conditions, frost resistant, tolerates prolonged snow cover, sensitive to shade and dry soil conditions; up to 2000m a.s.l., bi-annual to perennial.	Stem 0.2–0.7 m Root 0.2–0.8m, much branched	0.6–0.9	1100–1600
Trifolium pratense (Red clover)	*Arrhenatheretalia, Calthion, Molinion, Mesobromion, Eu–Nardion.* Fertile meadows and pastures, but also moist and poor meadows, sensitive in spring to grazing pressure, important feed plant; up to 2200m a.s.l., perennial.	Stem 0.2–1.2m Root up to 2m, much branched	1.5–2.3	450–670
Trifolium repens (White clover)	*Cynosurion, Plantaginetalia.* Heavily used turf, meadows, parks, aerodromes in humid areas, very prolific; up to 2300m a.s.l., perennial.	Stem 0.1–0.5m Root 0.7m, rooting at nodes	0.6–0.8	1250–1700
Vicia sativa (Common vetch)	*Brometea, Chenopodietea, Secalinetea.* Valuable cover crop; up to 1600m a.s.l., annual, many varieties.	Stem 0.3–1m, prone or climbing Root up to 0.5m	46	22
Vicia villosa (Fodder vetch)	*Papaveretum argemone, Secalinion.* Cover crop; up to 1700m a.s.l., frost hardy if sown early, annual to bi-annual.	Stem 0.3–0.6m, prone or climbing Root up to 0.6m	29	35
Conifers				
Larix decidua (European larch)	Constituent of coniferous forests, but also in mixed woodland, pure stands in certain localities in the Western and Southern Alps; main distribution in the alpine-continental spruce and arolla pine forests up to the tree line at 2100–2400m a.s.l.	Up to 35m Deciduous tap root system	5.9	170

Latin and English name	Plant association*, habitat and properties	Morphology: stem and root	1000 grain wt	Seeds/g
Conifers *continued*				
Picea abies (Norway spruce)	Pure and mixed stands on moist, humic, slightly acid soil derived from siliceous and calcareous rocks. *Vaccinio–Piceion, Eu–Fagion*. Optimum in the montane and subalpine region between 800–1900 m a.s.l.	Up to 30 (35) m, evergreen Dominant, widespread shallow roots, in very deep soil some vertical roots	7.7	130
Pinus sylvestris (Scots pine)	In mixed multi-species pine forests on base-rich calcareous soil (*Erico-Pinetum, Seslerio–Pinetum, Carici–Pinetum, Ononido–Pinetum*) and in acid pine forest on poor acid soil (*Calluno–Pinetum, Astragalo–Pinetum, Vaccinio–Pinetum*). Pure and mixed stands up to 1600 (1900) m a.s.l., drought resistant, frost hardy pioneer tree.	Up to 15 (20) m, evergreen Tap root	6.6	15
Pinus uncinata (Mountain pine)	Montane and subalpine forest on very shallow, stoney, calcareous soils. *Pinetum montanae*. Optimum in the Western Alps, in the northern limestone Alps of the east-Alpine region eastward to Berchtesgaden; Pyrenees, up to tree line at 2000–2400 m a.s.l., locally in small patches in montane and subalpine boggy areas; hardy pioneer.	Up to 15(20) m, evergreen Tap root system	6.6	15
Broadleaved trees				
Acer platanoides (Norway maple)	In mixed woodland from low lying areas up to 1100 m a.s.l., on moderately acid soils, immature soils.	Up to 30 m Deep rooted	125	8
Acer pseudoplatanus (Sycamore)	In moist, cool, mixed broadleaved tree woodland from peri-alpine hills to 1700 m a.s.l., requires humus-rich soil adequately supplied with water, tolerant to long periods of gravel aggradation.	Up to 25 m Deep rooted	83	12

Species	Site/ecology	Growth and root development		
Alnus glutinosa (Common alder)	Forest pioneer of fens and riverbanks. *Alnion glutinosae, Alno–Padion*. On acid, moist to wet soil up to 1500m a.s.l., pioneer on immature soil, nitrogen fixing.	Up to 20m Root development depending on habitat, but always deeper than *Alnus incana*	1.25	800
Alnus incana (Grey alder)	Pioneer on alluvial immature soils, riverbanks in the montane region. Forms single species woodland on *Alno–Padion, Alnetum incanae* on alluvial flood plains; up to 1400 (1600)m a.s.l., nitrogen fixing.	Up to 20m Shallow rooted	0.68	1470
Betula pendula (Silver birch)	Pioneer of most forest types in Central Europe, particularly on sandy and poor soil, sub-dominant in open spruce–oak–beech–alder woodland; up to 1700m a.s.l., particularly suitable for mine dumps.	Up to 15(20)m Intensive shallow root system	0.14	7140
Betula pubescens (Hairy birch)	Pioneer on acid, wet peaty soil and immature soils on siliceous parent material in humid areas; up to 1800m a.s.l.	Up to 12m Intensive shallow root system	0.26	3846
Fraxinus excelsior (Ash)	May form pure stands in hardwood forests and in gorges (*Alno–Padion, Fagion*). Formerly important browse tree, therefore widespread plantings up to 1400m a.s.l., sensitive to late frost, important soil stabilising properties.	Up to 35m Extensive, deep root system, tough roots	70	14
Fraxinus ornus (Manna ash)	In the warmer zones of broadleaved woodlands (*Orno–Ostryon*) in the Rhine valley, lower slopes of the southern Alps and Pannonian areas up to 800m a.s.l.	Up to 8m Deep rooted	65	15
Prunus avium (Wild cherry, gean)	Constituent of open broadleaved woodland (*Carpinion, Fagion, Ulmion*). At higher elevations only on forest margins; up to 1300m a.s.l., planted up to 1700m.	Up to 15m Deep rooted	166–200	5–6

Latin and English name	Plant association*, habitat and properties	Morphology: stem and root	1000 grain wt	Seeds/g
Broadleaved trees *continued*				
Prunus padus (Bird cherry)	Constituent of mixed broadleaved woodland (*Alno–Padion, Fagetalia*) at higher elevations. Intolerant to shading; from low lying areas up to 1700 m a.s.l., on rich and fertile soils, resistant to flooding and gravel aggradation.	Up to 15 m Extensive root system with tough roots	45.5	22
Sorbus aria (Common whitebeam)	Sunny positions on base-rich soils in broadleaved and coniferous woodland in warmer areas; (*Quercion pubescentis, Fagetalia, Berberidion*); from low lying areas up to 1500 m a.s.l.	Up to 12 m, reaches age of 200 years Deep rooted	3.75–6	167–267
Sorbus aucuparia (Rowan)	Wide-spread in almost all humid forest types; from low lying areas up to 1800 (2000)m a.s.l., tolerates light shade only, on any soil type.	Up to 15 m Deep rooted, on deep soil	2.5–3	330–400
Shrubs				
Acer campestre (Field maple)	Common in broadleaved forests and spinneys (*Carpinion, Ulmion, Cephalanthero–Fagion, Acerion, Quercion pubescentis, Berberidion*); up to 800 m a.s.l., tolerates only light shade.	Slow growing shrub to small tree, suitable for hedges, suckers freely Widespread tough root system	83	12
Alnus viridis (Green alder)	Forms closed formations on montane and subalpine alluvial flood plains and shelves; (*Adenostylion, Alnetum viridis*). On wet, lower slopes in snow-rich areas; from 500 to 1800 m a.s.l.	Up to 3 m Shallow rooted soil improver under prolonged snow cover, procumbent, impedes tree growth	0.55–0.62	1600–1800

Species	Ecology	Growth		
Amelanchier ovalis (Snowy mespilus)	Occasional in sunny oak and pinewoods (*Quercion pubescentis, Pinetum sylvestris, Berberidion*), on rocky, stony slopes; up to 1800 m a.s.l.	Up to 3 m, showy Roots in rock fissures, widespread root system	76.9–83.3	12–13
Colutea arborscens (Bladder senna)	In sunny, warm locations of oakwoods (*Quercion pubescentis, Lithospermo–Quercetum, Berberidion*) in the south; up to 800 m a.s.l., on any soil type.	Up to 3 m Deep rooted	1	1000
Cornus mas (Cornelian cherry)	Sunny, open woodland and woodland fringes (*Quercion pubescentis, Berberidion, Alno–Padion*); up to 600 m a.s.l., tolerates light shade, suitable hedge plant.	Up to 4 m high shrub or small tree to 6 m Widespread tough root system	166.6	6
Cornus sanguinea (Dogwood)	In sunny, open broadleaved mixed woodland, wood verges (*Prunetalia, Alno–Padion, Carpinion, Quercion pubescentis, Cephalanthero–Fagion*); up to 1000 m a.s.l., any soil type, light shade only.	Up to 3.5 (5) m, spreading suckers readily Widespread tough root system	30.3	33
Corylus avellana (Hazel)	Secondary formations in the temperate broadleaved mixed woodland zone (*Carpinion, Alno–Padion, Prunetalia, Querco–Fagetea*) in sunny, sub-mediterranean/sub-atlantic locations; up to 1400 m a.s.l., on many soil types, tolerates moderate shade.	Shrub or small tree to 5 (6.5) m, spreading habit Extensive tough root system	1000	1
Crataegus monogyna (Common hawthorn) *Crataegus oxyacantha* (Midland thorn)	Constituent of sub-atlantic and sub-mediterranean broadleaved and coniferous woodland (*Quercion, Carpinion, Pinion, Ulmion, Fagion*); up to 1000 m a.s.l., may be clipped annually, suitable for hedges.	Up to 6 (10) m, occasional as tree when trained, thorny Deep rooted Can reach 100 years of age	8.3	125
Cytisus scoparius (Broom)	On acid, non-calcareous soil on sunny slopes in areas of mild winters, (*Calluno–Sarothamnion, Sambuco–Salicion, Carpinion, Quercetalia roboris*); from low lying areas to 1100 m a.s.l., pioneer on immature soil, nitrogen fixing.	Up to 2 m. Deep extensive root system	8	120

Latin and English name	Plant association*, habitat and properties	Morphology: stem and root	1000 grain wt	Seeds/g
Shrubs *continued*				
Euonymus europaeus (Spindle tree)	Thickets and woodland fringes, open woodland (*Pinion, Alno–Padion, Carpinion, Fagion*); from low lying areas to 1100m a.s.l., on calcareous soil and sunny positions, indicates loamy soil, semi-shade.	Up to 4m shrub or small tree. Extensive root system	23–25	40–43
Frangula alnus (Alder buckthorn)	Common in bogs and damp woods, riverbanks, alluvial areas, open oak and pinewoods (*Alnion, Molinion, Alno–Padion, Quercion roboris, Luzulo–Fagion, Calluno–Genistion*); up to 1000 (1300)m a.s.l., tolerates flooding and oxygen-poor soil due to compaction.	2–7m, loosely erect, occasionally small tree. Shallow rooted with root bulbils	30	33
Genista germanica (German greenweed)	On acid soil of sunny slopes in the peri-alpine area (*Calluno–Genistion, Quercion roboris*); up to 750m a.s.l., indicates progressively acid soil conditions.	Up to 1.5m Spreading root system	3.2	312
Genista tinctoria (Dyer's greenweed)	Common on poor soils on sunny slopes in the peri-alpine areas, infertile meadows, oak forests (*Molinion, Calluno–Nardetum, Quercion roboris*); lowlands up to 750m a.s.l., indicates alternate wet–dry conditions.	Up to 1m Root up to 1m	3.3	300
Hippophae rhamnoides (Sea buckthorn)	Closed formations, pioneer vegetation on gravelly alluvia in the peri-alpine area, in dry pine forests on sandy soil and sand dunes of the North and Baltic Seas (*Berberidion, Erico–Pinion, Alnetum incanae*); up to 1000m a.s.l., forms root nodules, intolerant to shade.	Up to 3 (5)m, either small trees or thicket forming, suckers freely	7.5	133

Species	Habitat	Growth form		
Laburnum anagyoroides (Golden rain)	Constituent of sub-mediterranean oak woods and pine forests (*Quercion pubescentis*, *Lithospermo–Quercion*); up to 1000m a.s.l., soil improver, can be vegetatively propagated.	Large shrub to 7m, ascending branches Strong widespread root system	14–32	30–70
Laburnum alpinum (Alpine laburnum)	Constituent of montane and subalpine beech–fir tree woodland (*Fagion*) in humid areas, in the Alps replaces *Alnus viridis*, vegatively propagated.	Up to 5m, ascending branches	20–40	25–50
Ligustrum vulgare (Privet)	In sunny broadleaved and coniferous woodland on neutral- to base-rich soil (*Berberidion*, *Quercion pubescentis*, *Erico–Pinion*, *Alno–Padion*, *Carpinion*); pioneer plant up to 1000m a.s.l., vegetatively propagated, can be clipped (hedges).	Up to 2(3)m Extensive roots, runners	40	50
Lonicera xylosteum (Fly honeysuckle)	Common shrub in broadleaved, coniferous mixed woodland (*Fagion*, *Quercion pubescentis*, *Prunion*); from low lying areas to 1100(1600)m a.s.l., tolerates moderate salinity and shade.	Up to 2m loosely spreading Shallow rooted	10	100
Prunus mahaleb (St. Lucia cherry)	Sunny slopes and oak–pine forest in warmer areas (*Quercion pubescentis*, *Lithospermo–Quercetum*, *Berberidion*); up to 800m a.s.l., semi-shade.	Up to 4m loosely spreading, rarely a small tree Deep rooted	90–100	10–11
Prunus spinosa (Sloe, blackthorn)	In sunny positions, woodland fringes and open woodlands, (*Prunetalia*); up to 1000m a.s.l., pioneer on immature soils, intricately branched.	2–3m Creeping roots	1000	1
Rhamnus cathartica (Buckthorn)	Singly in wood verges and sunny spinneys, (*Prunetalia*, *Quercetum pubescentis*); mostly on calcareous soil; lowlands up to 1300m a.s.l.	2–3m, much branched, thorny shrub or small tree up to 6m, slow growing Extensive root system	14.3	70

Latin and English name	Plant association*, habitat and properties	Morphology: stem and root	1000 grain wt	Seeds/g
Shrubs *continued*				
Rosa canina (Dog rose)	Common in sunny positions, spinneys (*Prunetalia*); up to 1300 m a.s.l., on many soil types.	Up to 3 m, loosely spreading Deep rooted	2.8–3.3	30–35
Rosa rubiginosa (Sweet-briar)	Cosmopolitan in sunny positions, spinneys and woodland in warmer areas, prefers calcareous soils (*Berberidion, Prunetalia*); up to 1200 m a.s.l., indicates loamy soil.	Up to 3 m, loosely spreading, thorny Deep rooted	10	100
Rubus fruticosus (Blackberry)	Forest floor pioneer in humid areas with mild winters; lowlands up to 1600 m a.s.l.	Up to 1 m, arching branches, fast spreading Root shoots	2	500
Sambucus nigra (Elder)	Common in damp woodland, wasteland, spinneys, prefers damp and fertile soils (*Alno–Padion, Fagetalia, Sambuco–Salicion, Brometalia, Robinia* woodland); up to 1200 m a.s.l., indicates nitrogen-rich soil.	Up to 5 m, much branched, wide crown, shrub or small tree Shallow rooted	2.5	400
Sambucus racemosa (Alpine elder)	Often in damp and shady woods, cleared woodland, spinneys, in the montane region (*Fagion, Berberidion*); lowlands up to 1800 m a.s.l., usually on non-calcareous soil, nitrate indicator.	Up to 3 m, tall spreading bush Shallow rooted, new shoots produced from roots	7	143
Viburnum lantana (Wayfaring tree)	Scattered in open pine and oak woods, spinneys on calcareous soils (*Ligustro–Prunetum, Berberidion, Quercion pubescentis, Erico–Pinion*); up to 1400 m a.s.l., sprouting branches, tolerates severe cutting, hedge plant.	Up to 3 (4) m, spreading Extensive root system	43.4–45.4	22–23
Viburnum opulus (Guelder rose)	Common on alluvial shelves, bushes and broadleaved woods (*Prunetalia, Alno–Padion*), hydromorphic deep alluvial soil; up to 1000 m a.s.l., cut branches shoot readily, indicates moving water table.	Up to 5 m, fast growing, large bush or small tree Extensive shallow root system	40	25

Table 2.6 Vegetatively propagated trees and woody plants suitable for water bioengineering techniques.

Latin and English name	Size	Habitat	Vegetative propagation % take
Trees			
Populus nigra* (Black poplar)	Up to 30 m	Softwood alluvials, hydromorphic but well aerated sandy silty soils, periodically flooded; up to 1000 m a.s.l., in the southern Alpine region up to 1400 m a.s.l.	70–100, but only end cuttings with terminal bud, preferably from suckers
Salix alba* (White willow)	Up to 20 m	Softwood alluvials, lowland/lower montane regions, neutral fertile and calcareous alluvial sandy loams and loamy sands subject to periodic flooding, tolerates silt aggradation; up to 900 m a.s.l., in the Southern Alps to 1300 m a.s.l.	Approx. 70
Salix alba subsp. *vitellina* (Golden willow)	Up to 20 m	Cultivated willow with yellow or reddish-orange young twigs; only used as ornamental on site.	Approx. 70
Salix daphnoides* (Violet willow)	Up to 15 m	Softwood alluvials of mountain streams, particularly in the montane zone of the limestone Alps; on loamy, gravelly–sandy, neutral- to base-rich alluvial soils; up to 1300 m a.s.l., neutral central regions up to 1850 m a.s.l.	100
Salix fragilis* (Crack willow)	10–25 m	Permanently wet soils, poor base status, moving water table in areas with cool summers; alluvials in higher lying areas, can tolerate stagnant water table and gley conditions; up to 600 (1100) m a.s.l.	70–100
Salix pentandra* (Bay-leaved willow)	Up to 12 m	Coppices and alluvial woodland on wet, slightly acid alluvial soil of restricted permeability, mainly in lowlands and inner-alpine valleys; up to 1800 m a.s.l.	100
Salix rubens*	Up to 25 m	Softwood alluvials.	70–100

a.s.l. = above sea level.
*These plants are particularly suited for water engineering at lower and medium altitudes. Others are for use in the control of mountain streams and torrents.

Latin and English name	Size	Habitat	Vegetative propagation % take
Shrubs			
Laburnum alpinum (Alpine laburnum)	Up to 5m	In the Southern Alps in warm, moist woodland and bushes, on stony, rocky sites; 500–1900m a.s.l.	70–100
Laburnum anagyroides (Golden rain)	3–8m	Oak and pine woods in the Alps, mainly southern Alps on fertile and calcareous, humous, sandy to stony loamy soils in not too dry locations, mild winters.	70
*Ligustrum vulgare** (Privet)	Up to 3m	In warm broadleaved woods and pine woods, shrub associations; from lowland to 1000m a.s.l.	70–100
*Salix appendiculata** (Goat willow)	Up to 4 (6)m	From peri-alpine areas to the tree-line in humid locations on base-rich, neutral to slightly acid soil or calcareous rubble; 1200 (500)m to 2000 (2100)m a.s.l.	50–70, but strict adherence to winter dormancy
Salix aurita (Round-eared willow)	Up to 2.5m	From lowland to the montane regions, rare in areas of pronounced continental climate, marshes, on acid peaty gley soils; up to 1600m a.s.l.	50–70
*Salix cinerea** (Grey sallow)	2–3m	From coastal marshes to mountainous regions, in the warmer areas in drying marshes, fens; alder coppices, on fertile acid sandy soils to clay soils, tolerates waterlogged conditions (gley soils); up to 800m a.s.l.	50
*Salix eleagnos** (Hoary willow)	Up to 6m, rarely 15m	Characteristic constituent of *Hippophae rhamnoides–Alnus incana* woodland–bush, pioneer in the Erica-pine forest and grey alder alluvial woodland; mainly on alluvial shelves of alpine river valleys; calcareous rubble just above the ground water table, but periodically dry, sandy–stony slides on steep slopes.	50–70, but strict adherence to winter dormancy
Salix foetida	1.5m	Subalpine region of the central western Alps, in willow and green alder woods, riverbanks and wet slopes, on acid, fertile, often marshy soil, by preference on moraines of siliceous rocks; 1700–2000m a.s.l.	50–70, slow growing

Species	Size	Description	
Salix glabra	1.5 m	Limestone mountains of the eastern Alps, on rubble, stony slopes and gullies, only on calcareous and dolomitic rocks; 1400–2000 m a.s.l.	70–100, slow growing
Salix glaucosericea	Up to 1.5 m	Scattered in the subalpine region of the central Alps in green alder woods and scrub on wet siliceous moraines and alluvial shelves; 1700–2000 m a.s.l.	70
Salix hastata (Mountain willow)	Up to 3 m, rigid branches	In the high montane and subalpine regions of the Alps in willow–green alder scrub–bush areas in moist and shaded locations, neutral to weakly acid, fertile soils on various parent rocks, tolerates prolonged snow cover; 1600–2100 (2400) m a.s.l.	70–100
Salix hegetschweileri	Up to 4.5 m	Subalpine and montane region of the central Alps on wet, usually weakly calcareous but fertile soils subject to percolating water, riverbanks and wet lower slopes of bayleaf willow bush areas; 1600–2000 m a.s.l.	70–100
Salix helvetica (Swiss willow)	Up to 1.5 m	Subalpine region of the central Alps, on the shaded scree slopes with dwarf shrubs and green alder scrub, non calcareous, wet rubble, skeletal soils, tolerates prolonged and deep snow cover; 1700–2600 m a.s.l.	50–70, slow growing
Salix mielichhoferi	Up to 4 m	In the central part of the eastern Alps on wet slopes and riverbanks in the montane and subalpine zone on fertile skeletal soils; 1300–2200 m a.s.l.	70–100
*Salix nigricans** (Dark-leaved willow)	Up to 8 m bush or small tree	On moist, neutral to weakly acid clayey, gravelly or sandy soil, particularly in the cool–humid limestone areas; up to 1600 m a.s.l.	100

Latin and English name	Size	Habitat	Vegetative propagation % take
Shrubs *continued*			
*Salix nigricans subsp. alpicola**		The sub-species *alpicola* forms closed stands in the high-montane and subalpine regions of the central Alps, tolerates waterlogged conditions and shade.	
*Salix purpurea** (Purple osier)	Up to 6m	Softwood alluvial bush, often in the pioneer stage, periodically flooded, usually calcareous alluvial, silty, sandy, gravelly soils; up to 1600m (2300)m a.s.l.	100 most suitable willow for vegetative protection measures
*Salix triandra** (Almond-leaved willow)	2–4m	Softwood alluvial woodland, particularly in the pioneer stage, periodically flooded, wet, often calcareous, silty, sandy or gravelly soils; from lowlands to 1500m a.s.l., not very tolerant to shade, mainly in peri-alpine river valleys.	70–100
*Salix viminalis** (Common osier)	Up to 5m	Peri-alpine river valleys, often planted and therefore in the alpine river valleys; up to 1400m a.s.l., on intermittently wet, base-rich and fertile silty–loamy sands.	70–100
Salix waldsteiniana	Up to 1.5m	In subalpine zones of the eastern Alps on moist, neutral to weakly acid, base-rich, loamy, skeletal soils, tolerates prolonged and deep snow cover; 1400–2200m a.s.l.	70–100 slow growing

There is evidence that under certain, not sufficiently researched circumstances, *Salix caprea*, *Alnus incana* and *Alnus glutinosa* were used with varying success. In general, this cannot be recommended at the present time.
The bracketed numbers under 'Habitat' refer to higher elevations associated with the subalpine climatic regime.

Table 2.7 The most important rooted woody plants suitable for vegetative methods of construction.

Latin and English name	Elevation a.s.l.	Pioneer plant	Resistance to rockfall
Conifers			
Larix decidua (**European larch**)	foothills to subalpine, up to 2300 (2400)m	●	
Pinus sylvestris (**Scots pine**)	foothills to subalpine, up to 1600 (1900)m	●	
Pinus uncinata (Mountain pine)	submontane to subalpine, up to 2300 (2400)m	●	
Broadleaved trees			
*Acer campestre** (Field maple)	foothills to submontane, up to 1000m		
Acer platanoides (Norway maple)	foothills to submontane, up to 1100m		
Acer pseudoplatanus (Sycamore)	submontane to subalpine, up to 1700m		
Alnus glutinosa (**Common alder**)	foothills to submontane, up to 1050m	●	●
*Alnus incana** (**Grey alder**)	submontane to montane, 500–1600m	●	○
Betula pendula (Silver birch)	montane to subalpine, 1100–1800m	●	○
Betula pubescens (Hairy birch)	montane to subalpine, 1100–2100m	●	
Carpinus betulus (**Hornbeam**)	foothills to submontane, up to 1000m		○
Castanea sativa (Sweet chestnut)	foothills to submontane, up to 700 (1000)m		
Fraxinus excelsior (**Ash**)	foothills to montane, up to 1400m	●	○
Populus alba (**White poplar, abele**)	foothills to submontane, up to 800m	●	●
Populus nigra (**Black poplar**)	foothills to submontane, up to 800m	●	●
Populus tremula (**Aspen**)	submontane to montane, up to 1400m	●	○
Prunus avium (Wild cherry, gean)	foothills to submontane, up to 1300 (1700)m	●	
Prunus padus (Bird cherry)	submontane to subalpine, up to 1700m	●	
Quercus petraea (Durmast, sessile oak)	foothills to submontane, up to 1000m		○
Quercus robur (Common oak)	foothills to montane, up to 1200m	●	
Salix alba (**White willow**)	foothills to submontane, up to 900 (1300)m	●	●
*Salix caprea** (**Sallow**)	foothills to subalpine, up to 1700m	●	
Salix daphnoides (**Violet willow**)	submontane to subalpine, up to 1300 (1850)m	●	
Salix fragilis (**Crack willow**)	foothills to submontane, up to 1100m	●	
Salix pentandra (**Bay-leaved willow**)	foothills to subalpine, up to 1800m	●	●
Sorbus aria (Common whitebeam)	foothills to montane, up to 1500m	●	○

Latin and English name	Elevation a.s.l.	Pioneer plant	Resistance to rockfall
Broad leaved trees *continued*			
Sorbus aucuparia (Rowan)	foothills to subalpine, up to 1800 (2000)m		
Tilia cordata (Small-leaved lime)	foothills to montane, up to 1450m		
Ulmus glabra (Wych elm)	foothills to montane, up to 1400m		
Ulmus minor (Smooth-leaved elm)	foothills to submontane, up to 600m	●	
Shrubs			
Alnus incana (Grey alder)	submontane to montane, 500–1600m	●	
Alnus viridis (Green alder)	montane to subalpine, up to 1800 (2000)m	●	○
Berberis vulgaris (Common barberry)	foothills to submontane, up to 1800m	●	
Clematis vitalba (Traveller's joy)	foothills to submontane, up to 1000m		
Cornus mas (Cornelian cherry)	foothills to submontane, up to 600m		
Cornus sanguinea (Dogwood)	foothills to submontane, up to 1000m		
Corylus avellana (Hazel)	foothills to montane, up to 1400m		
Crataegus monogyna (Common hawthorn)	foothills to submontane, up to 1000m		
Euonymus europaeus (Spindle tree)	foothills to submontane, up to 1000m		
Hippophae rhamnoides (Sea buckthorn)	foothills to submontane, up to 1100m	●	
Laburnum alpinum (Alpine laburnum)	submontane to subalpine, up to 1900m	●	
Laburnum anagyroides (Golden rain)	foothills to submontane (subalpine), up to 1000 (2000)m	●	
Ligustrum vulgare (Privet)	foothills to submontane, up to 1000m	●	
Lonicera xylosteum (Fly honeysuckle)	foothills to montane, up to 1100 (1600)	●	
Pinus mugo (Dwarf mountain pine)	montane to subalpine, up to 1400 (1000) to 2300m	●	
Prunus spinosa (Sloe)	foothills to submontane, up to 1000m		
Rhamnus catharticus (Buckthorn)	foothills to montane, up to 1400m		
Ribes alpinum (Mountain currant)	submontane to subalpine, up to 1900m		
Ribes petraeum (Flowering currant)	submontane to subalpine, up to 1900m		
Rosa canina (Dog rose)	foothills to montane, up to 1350m		
Rosa rubiginosa (Sweet-briar)	foothills to montane, up to 1350m		
Salix appendiculata	montane to subalpine, up to 2000 (2100)m	●	

Species	Elevation	●	○
Salix aurita (Round-eared willow)	foothills to montane, up to 1600 m	●	
Salix caprea * (Sallow)	foothills to subalpine, up to 1700 m	●	
Salix cinerea (Grey sallow)	foothills to submontane, up to 800 m	●	
Salix eleagnos (Hoary willow)	submontane to montane (subalpine), up to 1400 (1850) m	●	○
Salix glabra	montane to subalpine, up to 2000 m	●	
Salix hastata (Round-eared willow)	montane to subalpine, up to 2100 m	●	
Salix hegetschweileri	montane to subalpine, up to 2000 m	●	
Salix nigricans (Dark-leaved sallow)	foothills to subalpine, up to 1600 (2400) m	●	○
Salix purpurea (Purple osier)	foothills to subalpine, up to 2300 m	●	
Salix repens (Creeping willow)	foothills to submontane, up to 1000 m	●	
Salix triandra (Almond-leaved willow)	foothills to montane, up to 1500 m	●	○
Salix viminalis (Common osier)	foothills to montane, up to 1400 m	●	○
Sambucus nigra (Elder)	foothills to montane, up to 1500 m		○
Sambucus racemosa (Alpine elder)	foothills to subalpine, up to 1800 m		○
Viburnum lantana (Wayfaring tree)	foothills to montane, up to 1400 m		
Viburnum opulus (Guelder rose)	foothills to submontane, up to 1000 m		○

Exotics for special use

Species	Elevation	●
Ailanthus altissima (Tree of heaven)	foothills, up to 500 m	●
Buddleia alternifolia (Buddleia)	foothills to montane, up to 800 m	●
Caragana arborescens (Caragana)	foothills to montane, up to 1000 m	●
Elaeagnus angustifolia (Elaeagnus)	foothills, up to 600 m	●
Forsythia intermedia (Forsythia)	foothills to montane, up to 1500 m	
Lycium barbarum (Duke of Argyll's Tea-tree)	foothills to montane, up to 1200 m	
Rhus typhina, Rhus laciniata (Sumach)	foothills to montane, up to 1000 m	●
Robinia pseudacacia (Robinia)	foothills to montane, up to 900 m	●
Rosa rugosa (Ramanas rose)	foothills to montane, up to 1000 m	●
Symphoricarpus racemosus (Snowberry)	foothills to montane, up to 1200 m	

a.s.l. = above sea level

Bold type indicates pronounced formation of adventitious buds and resistance to secondary covering by debris and/or soil.

● very tolerant to tolerant; ○ moderately tolerant

* Species that develop, depending upon environment, into either a shrub or a tree.

The bracketed numbers under 'Elevation' refer to higher elevations associated with the subalpine climatic regime.

Elevations: foothills to 500 m a.s.l.; montane 1100–1600 m a.s.l.; submontane 500–1100 m a.s.l.; a.s.l.; subalpine 1600–2300 m a.s.l.

Chapter 3

Water Bioengineering Systems and Techniques in Water Engineering

All construction approaches and techniques discussed in the companion publication dealing with ground bioengineering techniques for earthworks and erosion control (Schiechtl and Stern, 1996) are in principle applicable to the measures required for the protection of riverbanks and shorelines. The forces created by flow velocity and traction have, however, to be added to those of soil mechanics and soil movement. Several construction types that are specific to the protection of watercourses have to be added to the stabilisation and combined construction methods outlined in *Ground Bioengineering Techniques for Slope Protection and Erosion Control*.

The basic principle that vegetative means of construction can only rarely, if ever, replace the methods employed by classical civil engineering applies even more so to the protection of river courses. They maintain, however, their potential for use in combination with purely technical and mechanical methods of construction.

Schiechtl in 1973 introduced the following system of water bioengineering techniques:

(1) Soil protection techniques
(2) Ground stabilisation techniques
(3) Combined construction techniques
(4) Supplementary construction techniques.

Each of these methods and the relevant construction types have special functions and specific areas of application (Tables 3.1 and 3.2).

3.1 Soil protection techniques (Plates 2 and 3)

Soil protection techniques have the main objective of protecting the ground or soil surface: their effect does not extend into the subsoil. The

Table 3.1 Specific functions of bioengineering techniques.

Soil protection techniques protect the soil, by virtue of their dense surface cover, from sheet erosion and degradation. They improve the soil–water regime and soil temperature, thereby increasing the biological processes in the soil. Mulch cover affords the soil protection from raindrop action even before the sown grasses, herbs and woody species become established.

Ground stabilisation techniques modify or completely eliminate the mechanical forces active in the soil: they stabilise and protect lake shorelines, river banks, canal banks and natural slopes through the binding action of the root systems and increased evapotranspiration. They comprise linear or single point systems of bushes and trees, or their branches, which are capable of vegetative propagation, and are usually supplemented by the provision of a dense ground cover to guard against erosion.

Combined construction techniques support and stabilise unstable lake shore-lines, river and canal banks, in combination with solid structures (stone, wood, concrete, steel, plastics, etc.), resulting in increased effectiveness and pro-longed lifespan.

Supplementary construction techniques comprises seeding and planting designed to bring the project to its final shape and appearance.

large number of plants and plant materials per unit area will protect the soil surface from the deleterious effects of raindrop action and hail and from erosion caused by flowing water, wind and frost. The dense soil cover will improve soil temperature and water retention, thereby creating favourable conditions for plant development and growth in the topsoil and the air space immediately above ground level (Tables 3.1 above and 3.3. at the end of this chapter).

Table 3.2 Applicability of ground bioengineering techniques.

	Earthworks	Watercourse protection	Landscaping works
Soil protection	● ●	● ●	● ●
Ground stabilisation	● ●	○ ○	○
Combined construction	● ●	● ●	●
Supplementary construction	○	●	● ●

Key: ● ● very high;   ● high;   ○ significant;
    ● moderate;   ○ ○ low;   ○ very low

3.1.1 Turf establishment

Materials

Natural turf or sods that are available on site should be cut in thick sections complete with the rooted topsoil. Alternatively 'rolled turf', if required, is produced in special nurseries, reinforced with fibre mesh, which is cut and rolled up for easy transport and application. Its disadvantages are the rather limited species composition and thickness. It is therefore only suitable for favourable site conditions on moderately steep slopes with a well prepared seedbed of fine tilth.

Natural turves cut by hand are rarely larger than 400 × 400 mm. The size of mechanically cut turves depends upon the terrain and type of machinery used, but they rarely exceed 0.5 sq m.

Should there be an unavoidable time lapse between cutting and planting, the sods should be stored in clamps of maximum 1 m width and 0.6 m height to prevent desiccation and decomposition. Rodents such as field mice may cause considerable damage to such clamps during winter. Storage during summer is limited to a maximum period of four weeks. Valuable and sensitive turves for high altitude construction sites and nature preservation areas should be transported on pallets.

Rolled turf is made by seeding onto a thin bed of a granular substrate or on a large scale directly into a well prepared seedbed on deep soil. The former need only lifting from the beds, the latter are cut or peeled by hand or specially designed machinery. Such turf is available from nurseries in rolls 0.3–0.4 m wide and 1.5–2 m long; the thickness of the turf varies between 2.5 mm and a maximum of 40 mm. The rolls are transported on pallets and the maximum stack height should not exceed six rolls. Storage time including transport should not exceed four days and the stacks should be carefully protected from desiccation and bacterial heating. Rolled turf weighs 25–30 kg per sq m, depending upon the type of substratum used. A shrinkage loss of 5% may be expected on drying, which must be taken into account when ordering. It is feasible to grow rolled turf using special seed mixes to suit any given site conditions.

Implementation

Natural turf (Plate 2)   Natural turves may be used for the vegetative protection of slopes and watercourse banks by placing them very close to each other, taking care not to leave gaps. Planting in strips or in checkerboard fashion is not recommended, because the pattern is maintained for many years and is perceived as something artificial in the landscape. If there is only a limited amount of natural turf available, it is best used for the protection of the steepest slopes. On very steep slopes every fourth or fifth turf may be pegged to the ground using 300–500 mm long wooden stakes or metal staples. The pegs should not protrude above the turf surface to facilitate mowing. The use of live stakes for anchoring the turf is not recommended because rooting in the dense sward is very difficult. In addition, wire mesh, plastic grids or jute mats may be used on very steep and exposed slopes to secure the turf.

Rolled turf   Rolled turf in long sections is laid vertically down slope. On steep slopes it is necessary to secure the mats by driving stakes or pegs into the subsoil in a similar manner as indicated under the heading 'Natural turf'; the number of pegs per unit area may be reduced. Small sections of rolled turf are placed onto the soil surface the same way as natural turves. After placing onto the slope, the turf should be thoroughly tamped or rolled. The spreading of a thin layer of topsoil on the slope to be protected prior to the placement of rolled turf is not essential if it consists of grasses and plants specially selected for the prevailing site conditions.

Turf walls   Turf walls may be constructed of turves laid in courses to a height of 0.5 m. This method was very popular in bygone days, but is very rarely used today. To extend the life of turf walls, it is essential to reinforce them with steel pegs, wire mesh, geotextiles or geogrids.

3.1.2 Grass seeding

Many new methods of grass seeding have been developed during the past few years. Any of the methods described below makes it possible to establish a vigorous ground cover of grasses and herbs on sterile soils and soil-like material very quickly. The simultaneous seeding of grasses and woody species is a significant technical advance.

3.1.2.1 Hayseed seeding

Materials
0.5–2 kg of hayseed per sq m are required. Hayseed is obtained by collecting the sweepings from haylofts and barns. If the hayseed is to be used on its own, it is of advantage if a certain amount of stems and trash is part of the mix; 40–70 g of a compound fertiliser should be applied per sq m, depending upon site conditions.

Implementation
Hayseed is spread on the ground surface in a layer 30–40 mm thick. To avoid dispersion by wind, the soil should be moist or the seeds should be soaked in water before application. A combination of hayseed seeding with modern seeding methods is, however, more successful. This combines the advantages of using hayseed, i.e. a site-adapted species, with a mulch. At very high elevation in the Alps, hayseed will maintain its importance as long as it is available from the decreasing number of hay storage facilities.

Timing
Throughout the growing period, but for best results during its first third.

Effectiveness
The application of hayseed has a similar effect to mulch seeding, provided the mix contains a sufficient amount of stems and is spread in a thick layer. The ground cover protects the soil surface against mechanical forces and improves the micro-climate of the growing zones.

Advantages
Hayseed contains seeds of species that are not available commercially.

Disadvantages
Hayseed is only available in limited amounts and only in areas where natural meadows are still cut for hay.

Costs
Inexpensive. The seeding costs are relatively low at the equivalent of approximately 0.03–0.05 work hours per sq m, depending upon collection and transport. If hayseed is used in combination with other seeding methods the cost increases by 20–30%.

Areas of use
Hayseed is only used in areas where it is freely available and on sites where commercially available seed cannot be used. It is recommended for use in combination with mulch seeding and hydroseeding where multi-species plant communities are required.

3.1.2.2 Standard seeding

Materials
Legumes on their own or in combination with pioneer annuals to provide a cover crop and to prepare the project area for subsequent

afforestation or permanent cover; perennial grasses suitable for local conditions if a permanent cover is required.

Implementation
The seed is spread on the ground and lightly worked into the soil. The distance between the spreader and the ground surface should be as short as possible to prevent the possible separation of heavy and light seeds. Heavy and round seeds may roll downhill, collecting at flatter slope sections where they germinate, leaving the critical steep section bare. Very small and light seed should be mixed with dry sand or clay before application. Soil incorporation is best achieved by hand using a rake. Fully mechanised seeding is more cost effective on large areas, but limited to flat and even slopes or level ground. A large range of seeders and rollers is available for the purpose.

Standard seeding methods rarely provide satisfactory results on very poor soils or on subsoil. It is therefore preferable to restrict their application to areas where topsoil is available.

Timing
During the growing period; for best results at its start.

Effectiveness
No immediate effect. Only after germination, increased protection of the soil through the binding action of the roots and surface cover by the stems and leaves. Inoculated legumes will enrich the soil with nitrogen and organic matter, and the vigorous root system will improve the structure of the soil.

Advantages
A simple, quick and inexpensive method.

Disadvantages
Topsoil is required.

Costs
0.01–0.04 work hours per sq m; the lower value is applicable to flat areas and mechanised application, the higher value to hand sowing on steep slopes. The price of the seed mix may contribute significantly to total costs.

Areas of use
For the quick protection of temporarily disturbed areas and dumps to

prevent erosion by wind and water. Otherwise only as a green crop on rather level ground and gentle even slopes preparing the ground for permanent cultures.

3.1.2.3 Hydroseeding

Materials
1–30 litres sq m of ready seed mix which consists of seed, fertiliser, soil improvers, binding agents and water. The total volume required depends on local site conditions; site inspection and evaluation is essential for larger project areas.

Implementation
Seed fertiliser, soil improvers, binding agents and water are blended in a mechanical mixer to the consistency of a thin slurry which is sprayed with high pressure onto the ground surface. It is important that the mix is agitated vigorously in the tank during the operation. The aim should be to apply a uniform layer over the soil surface approximately 5–20 mm thick, which may be increased considerably on stony soil: two applications may be needed for layers exceeding 20 mm, allowing sufficient time between applications for the first to set.

Timing
Hydroseeding is only suitable for shady areas with high atmospheric humidity during the growing season. Hydroseeding with subsequent mulch cover is dealt with in Section 3.1.2.5.

Effectiveness
The blending of solids and fertilizers in the mix provides a good seed bed. The applied layer forms a soil-like crust which offers only limited resistance to mechanical forces, flowing water, frost and desiccation.

Advantages
May be used to seed very steep or stony sites and areas inaccessible to other methods and machines.

Disadvantages
Access to the area must be provided for the hydroseeder. The seed mix slurry may be sprayed to a distance of 40 m and, by using high pressure extension hoses, up to 150 m from the hydroseeder. Under normal circumstances the average machine may achieve a spray distance of 25 m.

Costs
0.05–0.1 work hours per sq m, depending upon the thickness of the applied mix.

Areas of use
Steep and stony, inaccessible ground, provided the machinery can reach the project area.

3.1.2.4 Dry seeding

In contrast to hydroseeding methods which use water as the medium to spread the seed, dry seeding is done by hand or by mechanical means, projecting or blowing the seed. Helicopters may be used to spread the seed mixed with the finely ground or granulated additives.

3.1.2.5 Mulch seeding

Mechanised mulch seeding

Materials
Seed; fertiliser; mulching materials such as straw, hay, cellulose or other fibres.

Implementation
A mulch spreader is used to blow a layer of chaff, shredded straw, etc. onto the seeded areas. Straw or hay is shredded in the mulch spreader, if required slightly wetted with a solution of an unstable bituminous emulsion to provide cohesion, and blown with great force onto the ground. A spray distance of 25 m may be achieved, 35 m with the use of extension pipes. In windy conditions it may be difficult to reach as far as 25 m.

Timing
Throughout the growing period.

Effectiveness
The applied chaff or straw binds together after the bitumen emulsion has set, providing a good soil cover where the layer is of sufficient thickness. The mulch influences the micro-climate by eliminating the extremes of temperature; however, the effectiveness required for extreme sites cannot be achieved owing to the relatively short length of the straw applied.

Advantages
Mechanised operation with high work output and low cost on average slopes and terrain.

Disadvantages
Only feasible in areas with good access, and of limited use on slopes steeper than 1:1 and under extreme site conditions. Limited spray reach. Uneconomical for use in small areas of less than one hectare due to the expensive machinery involved.

Areas of use
Low but long watercourse banks.

Mulch seeding using full-length straw (the Schiechteln method)

Materials
10–50 g/sq m of seed depending upon location and purpose; 300–700 g/ sq m of long straw or hay or synthetic fibres similar in structure; 40–60 g/ sq m of mineral fertiliser or 100–150 g/sq m of manure; 0.25 litres/sq m of stable bitumen emulsion; 0.25 litres/sq m of water; various biological or technical preparations depending upon site conditions.

Implementation
As a first step, long stem straw is spread on the ground resulting in a uniform cover; depending upon site conditions, the straw may first be subject to varying conditioning processes. In the second work phase, inoculated seed is sown, the mix consisting of species selected for local conditions, and fertiliser or manure. If required, other agents designed to improve and stabilise the soil together with growth promoting substances may be added. In the third phase, the straw mat is secured to guard against dislocation by spraying it with a special bitumen emulsion which does not harm the vegetation. In areas where the use of bitumen is either prohibited (water protection areas) or where the dark colour is not wanted, other binding agents may be employed. All these operations may be carried out by hand or mechanised means. The daily work output per gang varies from 3000 to 15 000 sq m depending upon site conditions. If the application of a binding agent to the mulch is considered to be insufficient for holding it together on steep or exposed sites, 350 mm long steel pegs may be driven into the soil at the rate of one per sq m. In addition, the pegs may be interlinked with wire to create an open mesh.

Timing
During the growing period.

Effectiveness
The mulch layer must be of a consistency that permits daylight to penetrate to the soil surface. The air space between mulch and ground surface must be large enough to create its own micro-climate which moderates extreme temperature fluctuations and permits the air to heat up during the day without reaching dangerous levels that could harm the growing plants. Moisture loss from the soil surface is reduced and radiation losses during the night cause condensation of water vapour. The physiological effects are very noticeable: on mulch-free control areas, plant growth is interrupted up to several times per day; under the mulch, such growing checks do not occur. As a result, plant production per unit time is increased to reach, under favourable conditions, a level comparable with the total production achieved during a full growing season by non-mulched areas. This is of particular importance at high elevation sites where the natural growing period is rather short; the dark colour of the bitumen plays a significant role in creating the necessary temperature conditions for rapid germination. The dark colour at other sites may be a disadvantage and the bitumen emulsion should be replaced by other paler binders.

Advantages
Simple, economical method that leads to quick success. Special access to the site is not required. With the exception of the bitumen spray, all other operations may be carried out by hand, although mechanical application is possible and many types of machinery are available. The method is particularly suited to difficult sites with infertile soils. The most labour intensive part of the method is the placing of the mulch layer which can be effected by the use of untrained labour.

Costs
Depend upon site conditions, access and type of vegetation cover to be established (cost of materials), they vary from 0.06 to 0.2 work hours per sq m.

Areas of use
Difficult terrain. On subsoils or infertile material.

3.1.3 Direct seeding of shrubs and trees

Two important areas of application may be distinguished:

❏ in areas that are inaccessible for traditional planting methods, such as stony and rocky slopes;

❏ as a supplementary measure at sites which have, after the completion of all bioengineering works, bare patches where revegetation measures were not successful.

Materials

Seeds of deciduous and coniferous trees and shrubs. Suitable seed varieties are listed in Table 2.5. The quantity of seed to be applied depends upon the seed size and germination percentage. The quality of the seed is governed by the rules and regulations laid down by the relevant authorities in each country. As is the case with plants and plant materials, special attention has to be paid to the origin of the seed. If the seed is to be spread on subsoil, inoculation with the appropriate symbiotic organisms is of advantage.

Implementation

Several methods established by silviculture are available:

❏ *Broadcast:* spreading the seed by hand or mechanically on the tilled or unprepared soil surface. This method lends itself particularly to the sowing of small seeds; hydroseeding is also possible (see Section 3.1.2.3), spraying a mix of seed (grasses, herbs, woody plants), fertilizers, additives and binding agent onto the soil.
❏ *Hole planting:* small holes of approximately 100 mm diameter and depth are dug with a hoe. 1–5 large seeds, e.g. acorns, and a pinch of the smaller seeds are inserted and covered with 10–20 mm of soil.
❏ *Spot seeding:* the seed is spread on small selected patches ranging in size from several sq cm to 1 sq m. Any existing vegetation is chopped down, the soil loosened and the seed applied.
❏ *Drilled seed:* the seed is placed by hand or mechanically into small and narrow furrows and covered with soil.

Timing

For best results seeding should take place in spring or autumn. Spring seeding is of advantage because the melting snow and/or winter rains will usually have brought the soil to full field capacity. Autumn seeding corresponds to the natural cycle and the seeds lie dormant over winter. This, in comparison to the abundance of nature, small amount of placed seed must be protected from predation by small animals and be treated with repellents. In the past, snow seeding was widely practised (spreading the seed on the snow) for the propagation of European larch (*Larix decidua*), birch (*Betula alba*) or alder (*Alnus glutinosa*).

Effectiveness
Under the shelter of existing vegetation the shrub and tree seeds will develop in a similar manner to the natural process of forest rejuvenation. Species diversity is increased and with it the number of plant communities.

Advantages
Simple and inexpensive method. No disturbance of the soil, thus preventing erosion. The great number and variety of seeds will ensure survival in the natural selection process.

Disadvantages
The procurement of seed suitable for the prevailing local conditions may be difficult.

Areas of use
No other seeding method can replace the seeding of woody plants on rocky soil and precipitous sites. On account of the economics involved and good results obtained with regard to diversity and root penetration, the seeding of woody species in extreme locations should be given preference over other bioengineering techniques.

3.1.4 Erosion control netting (Plates 4 and 5)

All the described seeding methods may be reinforced by the provision of netting or mesh made of jute, coir fibre, synthetic fibres or wire. Such netting will anchor the mulch layer and afford the topsoil extra protection. Due to the high costs of the netting, special reasons for its use must be apparent to justify the expense. Areas of application include slopes and watercourse banks with very sandy soils and steep sites subject to wind erosion. Netting applied to riverbanks should be anchored very thoroughly, as flood water may carry the net complete with vegetative cover with it (Plate 5).

3.1.5 Seed mats

Materials
Finished mats of variable construction which consist of natural or synthetic fibres, made up of two outer fibrous layers and an inner fabric for reinforcement.

Implementation
Seed mats should only be placed on moist soil with a firm tilth; gravels and other coarse grained material must first be covered with a layer of fine textured soil. Scarification of the soil is not necessary. After placing the mats, they should be rolled or pressed down to establish close soil contact. To prevent the mats from shifting they should be pegged down using stakes or steel staples, or alternatively have all edges buried to a depth of approximately 300 mm.

Timing
During the growing period.

Effectiveness
As the fibres decompose very slowly, long lasting and effective soil cover is achieved. They afford only limited protection against mechanical forces and can absorb only a limited amount of moisture.

Advantages
Long lasting protection. Effective prevention of surface erosion if not undercut by running water.

Disadvantages
Only suitable for flat areas with a fine tilth. Only standard seed mixes are available. Special mix requirements must be ordered well in advance of the expected installation period.

Costs
Depend to a large degree on the terrain and the make of the mat. Usually higher than other seeding methods.

Areas of use
Predominantly for grassed waterways.

3.1.6 Precast concrete cellular blocks

Materials
Cellular or perforated or voided concrete blocks, not reinforced, are available in many types and under a range of proprietary names; staples or steel pegs to anchor the blocks, one per sq m; topsoil; seed.

Implementation
Cellular concrete blocks are placed on the surface of the slope and secured with steel pegs or staples (approximately 1 peg per sq m). The

hollow spaces of the blocks are filled with topsoil and sown to a predetermined seed mix – this should be drought and alkali tolerant.

Effectiveness
Provides very good and long lasting soil cover. The final appearance of the vegetation cover depends on the type, shape and concrete to void ratio of the blocks.

Advantages
Immediate stabilising effect. Manufactured blocks freely available from commercial suppliers.

Disadvantages
Only a few of the presently manufactured blocks are designed for slope protection; the concrete to void or soil ratio is unfavourable, leaving too much unsightly concrete exposed. High costs.

Areas of use
Protection of the lower sections of unstable slopes where the blocks rest on sound foundations. Protection of riverbanks.

3.1.7 Live brush mats (Fig. 3.1; Plates 6–8)

Materials
Preferably straight branches of brush willows. Depending upon the amount of laterals, 20–50 branches or stems are placed per linear m if the slope length is equal to the length of the branches, which should not be less than 1.5 m. If the slope is longer, several rows of brush are constructed with an overlap of at least 300 mm. If there is a shortage of branches capable of vegetative propagation, a mix of non-shooting branches is permissible, aiming at a uniform distribution of the two types of material to ensure uniform shoot production.

Implementation
The branches should be placed on the ground aiming for 80% ground cover. The butt ends of the branches must be covered in soil to facilitate root formation. The ends of the branch layer closest to the water's edge are, in addition, secured by stones, poles, fascines or wattle fences. The construction of the mat proceeds from the top of the slope downwards so that the tops of the lower run overlap the butts of the upper run by at least 0.3 m. The mat is anchored in lines 0.8–1 m apart, using wire, stems,

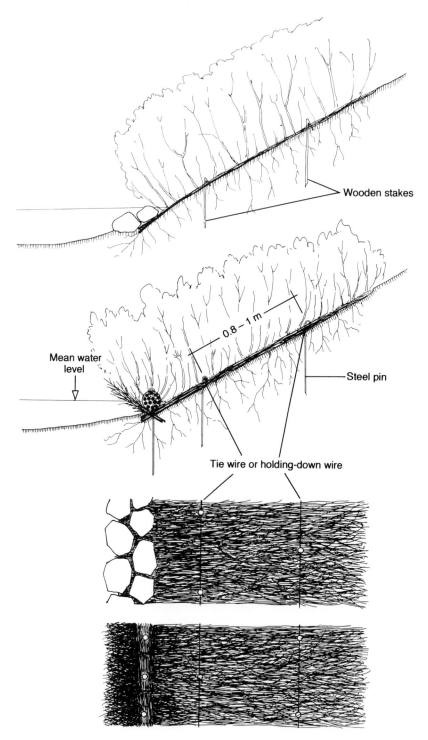

Fig. 3.1 Section and plan details of brush mats. *Top diagram in each pair*: branches secured with rip-rap. *Bottom diagram in each pair*: branches secured with live fascines.

fascines or wattle fences; the most simple method is to use wire. Wooden poles or notched steel pegs are driven through the mat at 0.75 m spacing and interlinked with wire; the pegs are subsequently driven into the ground to establish close contact of the mat with the soil. To protect the recently established mat from flood water damage, netting or similar devices should be placed over the branches at exposed positions. To facilitate rooting, the mat must lie snug on the well prepared smooth ground, and it should be covered lightly with topsoil.

Brush mats may be prepared in advance by tying slim branches and stems with wire or polypropylene twine for transport in roll format to the site for placement (Karl, 1990).

Timing
Only during the dormancy period.

Effectiveness
The mats provide effective cover after completion of works, protecting the slope from surface erosion. Intensive shoot and root development occurs in a short space of time.

Advantages
Very resistant to tractive forces. According to Florineth (1982) tractive forces of 200 N/sq m after completion, 300 N/sq m after 15 months and 400 N/sq m after the third year can be withstood.

Disadvantages
Requires vast amounts of vegetative material. Labour intensive method.

Costs
Depending upon the slope conditions and procurement of plant material, 1–5 work hours/sq m.

Areas of use
For the speedy and long lasting protection of riverbanks liable to flood damage.

According to its method of construction, the brush mat technique may be classified as 'cover works'; from the materials' point of view, it could be grouped with ground 'stabilisation works'. It is as such the link between the two methods.

3.2 Ground stabilisation techniques

(Figs 3.2–3.5; Plates 9–12; Table 3.4 near the end of this chapter)

Ground stabilisation techniques are required where harmful mechanical forces occur within the subsoil. The effectiveness of the described measures depends upon the depth to which they penetrate, and upon the distance between them. With increasing root penetration, stabilisation increases, reaching a maximum after several years when the established vegetation cover reaches maturity. Since stabilisation structures are generally linear or pinpoint measures, a good ground cover should be established at the same time.

Ground stabilisation is mainly concerned with earthworks and the subject is treated in greater detail in *Ground Bioengineering Techniques for Slope Protection and Erosion Control* (Schiechtl and Stern, 1996). Its role in steam bank protection (water engineering) is confined to preventing soil movement, erosion of riverbanks and eliminating slippage zones; the live brush layer method, one of the classic measures for the extensive protection of riverbanks against erosion, has so far only been accepted with some hesitation, although the construction period, usually confined to the dormancy period, could be extended into the growing season by storing the live material in flowing water. Moreover, live materials can be cut during the growing period for immediate use in permanently moist habitats (see Sections 2.3.3 and 2.5.1).

3.2.1 Live cuttings (Figs 3.2 and 3.3; Plate 9)

Materials
Cuttings are unbranched, one to several years old shoots of suitable shrub and tree species, 10–50 mm in diameter with a minimum length of 400 mm (see Table 2.6).

Implementation
The shoots, their lower ends cut obliquely, are driven vertically or at an angle and by hand, if necessary using a mallet, into the soil; if automatic jack hammers are used for the purpose, a piece of closed metal tube corresponding to the diameter of the cutting is placed over its top to prevent it from splitting. In very dense soil it may be necessary to prepare the planting holes with a metal rod or drill. The cutting should protrude only to a maximum of one-quarter of its length above ground level to prevent it from drying. The cuttings should not be placed in rows or at regular intervals, but at random in the most suitable places at a rate of 2–5 cuttings/sq m.

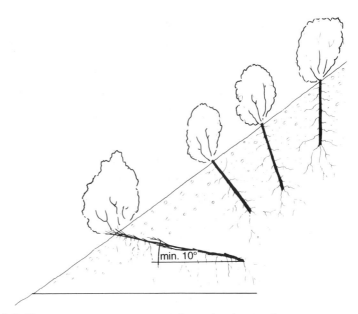

Fig. 3.2 Planting cuttings at various angles to the slope surface.

Timing
Dormancy period; if stored in running water, until the early growing season.

Effectiveness
Soil stabilisation commences only after rooting.

Advantages
Quick technique.

Costs
Very low, output 2–5 sq m per work hour, including all preparations.

Areas of use
All types of bioengineering techniques.

3.2.2 Wattle fence and wattles (Figs 3.4 and 3.5; Plate 10)

Materials
Long, pliable stems of vegetatively propagating shrub and tree species.

Implementation
Wooden stakes of 30–100 mm diameter and 1 m length or corresponding steel rods are hammered at 1 m intervals into the soil. Shorter stakes or

Fig. 3.3 Riverbank planted to willow cuttings. *Above*: after completion of planting. *Below*: after 18 years.

Fig. 3.4 Willow wattle fence and braided groyne five years old in a mountain stream.

rods or live stakes are inserted between the main stakes or rods at 500 mm intervals. Strong and flexible stems capable of vegetative propagation are then plaited between the stakes. Each pair of stems, after plaiting, is pressed firmly down and a total of 3–7 pairs are placed one on top of the other. Alternatively, previously prepared wattles may be secured to the stakes. The stakes should not protrude more than 50 mm above the wattles and at least two-thirds of their length should be firmly embedded in the soil. The lowest stem and the cut ends of all others must be buried in the soil to allow rooting. Wattle fences placed totally below surface level have a better chance of rooting. The stems above ground level are likely to dry and perish. Wattle fences may be arranged in rows along contours or in a diagonal pattern, forming a grid across the slope.

Timing
Only during the dormancy period.

Advantages
Immediate stabilising effect; enable stepped slopes and banks to be formed.

Fig. 3.5 Honeycombed wattle system with braided willows immediately after completion.

Disadvantages
Large quantities of live plant material needed. Relatively low rooting percentage because many stems do not take readily (especially if exposed by seepage slumping). Only long and flexible stem cuttings may be used, excluding many valuable willow species, particularly in the Alps. High costs and labour input.

Costs
0.8–1.5 work hours per linear m.

Areas of use
Wattle fences should always be buried in the soil. They fulfil a limited but important function, such as the immediate protection of smaller slippages and the retention of the topsoil, provided its depth does not exceed 100 mm, and in combination with other construction methods, for the protection of riverbanks and drainage ditches.

Apart from the wattle fence, other braided structures such as the 'reinforced willow plait' are in common use for the protection of riverbanks. Other labour intensive methods that require large amounts of vegetative materials and which were formerly much used in Germany for the protection of very steep riverbanks include an intensive woven arrangement of honey-comb willow anchored with live willow poles, and the so-called Fischer pole fence which was used extensively for the erosion control of the River Enns in Austria. This type of fence is a massive structure, protruding 0.5 m or more above the ground level. To protect the braided willows from being undercut by flowing water, they are placed on a brush layer which on the land side is secured with large stones. On the river side, the branches of the brush layer must protrude for 0.5–0.75 m beyond the fence. The Fischer fence protects alluvial land from floating debris but permits reasonable over-topping by flood waters.

3.2.3 Layer construction (Plates 11 and 12; Table 9)

The main types of layer construction can be distinguished:

❏ hedge layer using rooted plants;
❏ brush layer using branches of woody plants capable of vegetative propagation;
❏ hedge–brush layer using both rooted plants and branches of woody plants.

The construction of any type of layer for the general protection of slopes or watercourse banks requires the following preparatory works. Small terraces or berms are cut into the slope with a base of 0.5–2 m width; on a very steel slope narrow trenches are excavated to a depth of 0.5 metres. Construction of the layers starts at the bottom of the slope, and the berm, after placing the branches, is backfilled with the soil from the next berm directly above. The base of the berm should slope at an angle of 5–10° backwards into the watercourse bank or cut. The terraces are arranged on the contour, or, if surface run-off or seepage water is to be disposed of, at an angle of up to 60° to the contour. Steeper angles will complicate construction. The distance between the berms depends upon the slope gradient and soil properties, and varies as a rule between 1 and 3 m. Closer spacing than 1 m is not advisable because of the danger of the top trench or berm collapsing. The necessary earthworks may be carried out by hand, or by using special back hoes and small crawler equipment.

3.2.3.1 Hedge layer

Materials
Rooted plants or saplings of woody deciduous plants which tolerate complete soil cover by producing adventitious buds and roots (see Table 2.7). By preference two- to three-year-old saplings are used; fast growing species such as alder may be used after two years. The stem to root ratio is of importance; plants with massive root systems will take much quicker and grow more vigorously. Depending upon species, five to twenty plants per linear m are required.

Implementation
Preparation of berms 0.5–0.75 m wide. To improve the soil, a thin layer of topsoil, compost or other organic manure may be spread on the base of the berm. The rooted plants are placed very close to each other, letting one-third of their length protrude beyond the slope surface.

Timing
Spring or autumn for open rooted plants; growing period for container plants.

Effectiveness
A deciduous mixed woodland may be established without the prior planting of a cover crop, provided the right species for the intended climax vegetation are selected.

Advantages
Shortening of the time normally involved for plant succession.

Disadvantages
Large quantities of plants needed. Only suitable for reasonable soil conditions and locations.

Costs
1–3 work hours per linear m.

Areas of use
On rather fertile soil and in areas where willows are scarce or not naturally growing.

Plate 1
Intensive new root
formation on a mature
white willow (*Salix alba*)
of 400 mm breast height
diameter, after one year
of artificial landfill.
(Courtesy Gall, Kufstein,
Austria)

Plate 2
Collection of natural turf from a natural habitat

Plate 3
Bank protection and shaping of an intermediate power
station reservoir three years after planting

Plate 4
Placing erosion control netting. (Courtesy Zeh, Worb, Switzerland)

Plate 5
Insufficiently secured erosion control netting displaced by high flood water

Plate 6
Brush mat in a torrent after the completion of works. (Courtesy Florineth, Schlanders, Bolzano, Italy)

Plate 7
Brush mat two years after establishment (compare with Plate 6)

Plate 8
Willow brush mat 20 years after establishment

Plate 9
Riverbank protection with live stakes or truncheons

Plate 10
Wattle fence in combination with a spare willow brush mat after completion

Plate 11
Steep riverbank reinforced by geogrids in combination with brush layer method. (Courtesy Sotir, Marietta, Georgia, USA)

Plate 12
Brush layer construction on a steep riverbank (compare with Plate 11) after completion of the lowest layer. (Courtesy Sotir)

Plate 13
Live palisade in a narrow river channel one year after completion

Plate 14
Live brush sill on a wide flood control berm

Plate 15
Construction of a live wire mesh gabion sill

Plate 16
Vegetatively reinforced live wire mesh gabion sill seven years after construction

Plate 17
(*below left*) Simple timber sill gully control structure

Plate 18
(*below middle*) Medium size timber cribs in gully and torrent control works

Plate 19
(*above right*) Large stone filled timber crib sill in a torrent with permanent flow

Plate 20
Well-established timber crib walls five years after construction

Plate 21
Wire gabion barrier during construction

Plate 22
Dry stone wall torrent
control structure near the
tree-line

Plate 23
Small groynes in a
mountain stream to
increase biodiversity

Plate 24
Vegetative reinforced groynes after completion of works. (Courtesy Sotir, Marietta, Georgia, USA)

Plate 25
Comparison with Plate 24, only two months after completion. (Courtesy Sotir)

Plate 26
Construction of a live groyne with wattles and brush layers. (Courtesy Zeh, Worb, Switzerland)

Plate 27
Groynes in a large mountain river, 100 years after completion

Plate 28
The construction of long groynes led to the establishment of alluvium areas only a few years after completion of the works

Plate 29
Still-water areas between groynes

Plate 30
Groynes in a mountain river with heavy bed load 100 years after completion

Plate 31
Short, groyne-like stone ribs caused very turbulent flow during flood periods, leading to severe damage of the bank between the ribs. (Courtesy Florineth, Schlanders, Bolzano, Italy)

Plate 32
Live brush traverses after completion of works

Plate 33
Rehabilitation of bank slump with a brush grid: first phase of construction

Plate 34
Brush grid before applied earth cover

Plate 35
Establishing reed culm plantings

Plate 36
Reed culm planting two years after establishment. (Courtesy Begemann, Lennestadt, Germany)

Plate 37
Reed clump planting eight years after completion

Plate 38
(*below left*) Reed roll construction in a lowland river

Plate 39
(*below right*) Comparison with Plate 38, five months after establishment of reed roll

Plate 40
Willow cuttings in stone revetment after one year

Plate 41
Willow cuttings in stone revetment after three years

Plate 42
Willow cuttings in stone pitching joints after 20 years in the lower reaches of a torrent

Plate 43
Tree growth after 40 years resulting from joint planting with cuttings

Plate 44
Construction of a bank fascine

Plate 45
Bank fascine secured with jute netting to prevent undercutting and wash. (Courtesy Sotir, Marietta, Georgia, USA)

Plate 46
Branch packing – classic construction with fascines. (Courtesy Sotir)

Plate 47
Construction phase of a massive live timber crib wall revetment. (Courtesy Sotir)

Plate 48
Close-up of construction of a massive live timber crib wall. (Courtesy Sotir)

Plate 49
Development of the live vegetative components in the crib wall revetment shown in Plate 47 after two years. (Courtesy Sotir)

Plate 50
Construction of a massive live timber frame on a steep bank. (Courtesy Zeh, Worb, Switzerland)

Plate 51
Construction of a brush-layered geogrid reinforced stream bank structure. (Courtesy Sotir, Marietta, Georgia, USA)

Plate 52
Brush-layered geogrid
reinforced floodbank
structure after
completion of works.
(Courtesy Zeh)

Plate 53
Detail of a flexible wire
rope-tied boulder bank
protection structure one
year after completion

Plate 54
Shrub growth resulting from a flexible wire rope-tied
riverbank protection installation on the lower reaches of
a mountain stream after ten years

Plate 55
Vegetation covered riverbank as a result of transplantation in a drainage ditch after three years

Plate 56
Lake shore double-row palisade wave barrier during construction. (Courtesy Zeh, Worb, Switzerland)

Plate 57
Trees along a river berm being incorporated into a floodbank. The fill was raised in stages to ensure the survival of the trees

Plate 58
The same trees after three years, covered to a height of 3 m with fill material

Plate 59
Retention of a group of trees of white willow (*Salix alba*), grey alder (*Alnus incana*) and bird cherry (*Prunus padus*) in the basin of a reservoir by raising the soil mantle in stages over several years

Plate 60
Comparison with Plate 59 after the completion of the cover, with the dam slowly filling

Plate 61
A group of white willows (*Salix alba*) 50–70 years old, 4 years after raising the soil level by 6 m!

Plate 62
The use of live building materials in the reconstruction of a large drain has created new wetland habitats. (Courtesy Göldi, Zürich, Switzerland)

Plate 63
Medium- to long-term development of riparian woodland for flood protection through the use of water bioengineering techniques

3.2.3.2 Brush layer

Materials
Branches of bushes and trees capable of vegetative propagation, essentially willows (see Table 2.6). Twenty branches per linear m.

Implementation
The branches are placed in a criss-cross manner on the base of the berm which is 0.75–2 m wide. Care should be taken to achieve a uniform mix of the species involved and of the various age groups. The branches should not protrude more than 0.2 m beyond the slope surface.

Timing
Dormancy period to early growing season if branches are stored in flowing water.

Effectiveness
Immediately effective after construction, stabilising action extending to the subsoil level. Little chance of flood damage as the layers are deeply covered in soil, with the angle of alignment of the branches affording added protection.

Advantages
Relatively simple construction. Fast formation of an effective root–soil matrix. Use of spreading branches of dwarf type montane willow varieties in the sub-alpine region possible.

Disadvantages
Not suitable for the stabilisation of very thick organic soil horizons.

Costs
0.7–2 work hours per linear m.

Areas of use
Protection of slopes prone to slides and of riverbanks. Rehabilitation of slips and slumps. (May be used in combination with synthetic or biodegradable geogrids or geotextiles, e.g. Plate 11.)

3.2.3.3 Hedge–brush layer

Materials
Vegetatively propagating branches and open rooted plants used in combination; ten branches and one to five rooted saplings per linear m.

Implementation
Branches and saplings placed alternatively onto the prepared berm so that branches and rooted plants protrude 0.2–0.3 m beyond the slope surface after backfilling the berm.

Timing
Dormancy period; early and late growing period.

Effectiveness
The use of rooted plants shortens the time span of natural plant succession and requires less time for establishing itself than brush layers (Schütz, 1989).

Advantages
Various stages of plant succession established in one operation.

Disadvantages
Limited application for the stabilisation of humus-rich topsoil.

Costs
0.8–2.5 work hours per linear m.

Areas of use
Wider areas of use in the erosion control and protection of watercourse banks and riverbanks as compared with brush layers. Suitable for all climatic zones where the plant material occurs naturally.

3.3 Combined construction techniques
(Figs 3.6–3.31; Plates 13–55; Table 3.5 near the end of this chapter)

Structures for the protection of riverbanks and riverbeds, for gully control and the erosion control of unstable slopes are not necessarily entirely of a mechanical nature ('hard or inert structures'), but may be combined to advantage with vegetative ('natural' or soft) measures.

Because combined construction techniques consist of live and inert materials, they are immediately effective. The effectiveness of the planted vegetation increases with its advancing age and size.

(I) Transverse structures

3.3.1 Bank to bank structures

Confined to narrow streams and gullies with small or intermittent flow, they are designed for consolidation of the bed and/or retention of the sediment load.

3.3.1.1 Gully control with brush work: branch packing (Fig. 3.6)

Materials
Live branches capable of vegetative propagation; poles; wire.

Implementation
In order to achieve a dense ground cover and to ensure the formation of an extensive root system, live branches are placed in herringbone fashion in the gully, the tips of the branches pointing outwards. The lower ends of the branches are progressively covered with earth up to a maximum depth of 0.5 m. To anchor the branch layer, sturdy poles are placed every 2 m across the gully, burying the pole ends firmly in the banks. Such structures tolerate periodic sediment cover or alternate conditions of sedimentation and erosion very well.

Timing
Only during the dormancy period.

Effectiveness
The intensive root penetration of the soil protects and stabilises the flanks and bottom of the gully.

Advantages
The use of live materials ensures a long lasting effectiveness.

Disadvantages
Large quantities of branches required.

Costs
Relatively inexpensive provided the branches are available from locations close to the construction site.

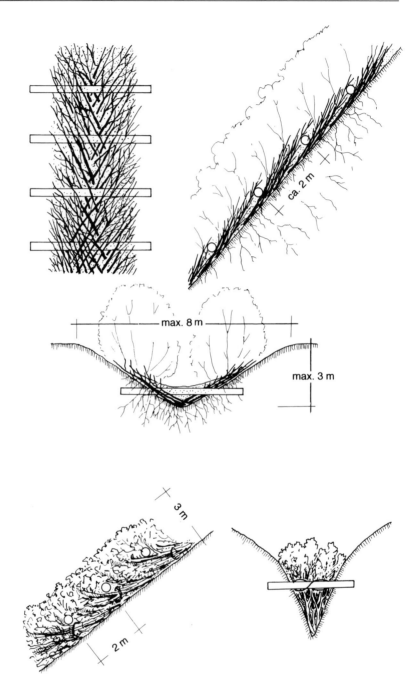

Fig. 3.6 Construction methods for gully control measures. *Above*: branch layers of woody species capable of vegetative propagation. *Below*: branch packing using non-rooting branches and stems of deciduous plants, and also branches of conifers.

Areas of use
Rehabilitation of gullies up to 3 m deep or with intermittent flow. Very successful in combatting persistent erosion.

3.3.1.2 *Palisades* (Fig. 3.7; Plate 13)

Materials
Fairly strong and straight branches capable of root formation; willow stems; poles; steel wire.

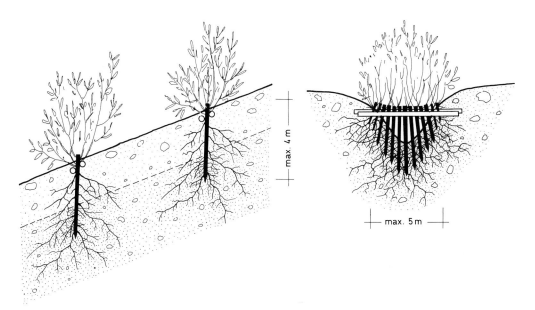

Fig. 3.7 Construction of live palisades using willow poles.

Implementation
Strong, live poles of uniform diameter, square cut at the top and pointed at the base, are hammered for one-third of their length into the ground at very close spacing and securely tied with rust-resistant steel wire or willow stems to cross beams anchored in the gully sides.

Timing
Only during the dormancy period.

Effectiveness
Immediately effective as a barrier even before taking root, causing siltation and bed elevation.

Advantages
Quick method. Has an immediate effect. Takes root easily. Simple and effective for the control of steep gullies by creating level sections through silt accumulation, and drop structures.

Disadvantages
Maximum width of structure 5 m and limited height of 2–4 m. Strong, straight poles only available in areas with good growing conditions.

Costs
Relatively low if the poles are locally available.

Areas of use
Creation of a stepped profile in steep gullies in climatically favourable, low lying areas. Particularly suitable for use in rather loose, fine textured soils such as loams, loess and sandy material.

3.3.1.3 *Live brush sills or submerged weirs* (Figs 3.8–3.13; Plates 14–16)

The method used in the construction of live riverbed sills or submerged weirs depends on the prevailing local conditions and the objective, i.e. bed erosion control, aggradation and base load retention. Very steep gullies require closer spacings of the structures, leading ultimately to the introduction of a stepped or terraced gully bed.

Brush sills (Fig. 3.8; Plate 14)

The brush sill is the most simple of all the live transverse structures.

Materials
Strong willow branches 1.5–2 m long; large sized crushed stone/rip-rap or alternatively cylindrical wire gabions; poles; fascines; stakes.

Implementation
A trench, triangular in transverse section, is dug across the bed and a thick layer of willow branches placed on the downstream side of the trench, so that two-thirds to one-half of the branches remain above bed level. The brush layer is anchored by placing crushed stone/rip-rap or cylindrical gabions, or timber poles or fascines across the upstream end, finally backfilling with the excavated material to the level of the bed: only the flexible ends of the branches must protrude. If crushed stone is freely available, the sill may be constructed of hand-placed stone and by

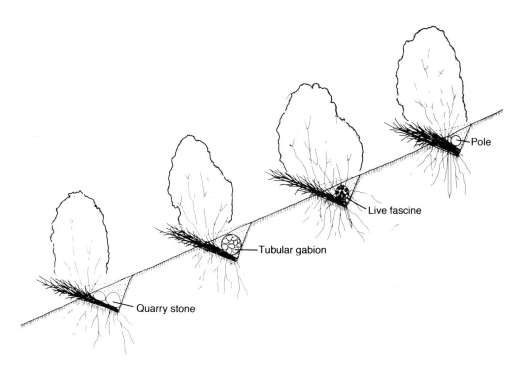

Fig. 3.8 Various types of brush sills.

inserting live willow cuttings in the joints (Fig. 3.9). Each riverbank, at its intersection with the sill, needs extra protection to prevent lateral erosion and undercutting, by construction pole deflectors tightly packed with willow branches and secured with steel wire; care must be taken not to leave any gaps in the transition zone between sill and deflector.

Timing
During the dormancy period.

Effectiveness
The sills prevent bed scour. The new shoots developing from the branch layer will cause sedimentation. Part of the branches will be pressed against the bed by the flow, preventing scour.

Advantages
Simple and quick method.

Disadvantages
Application is limited to small streams with very low or intermittent flow and low gradient.

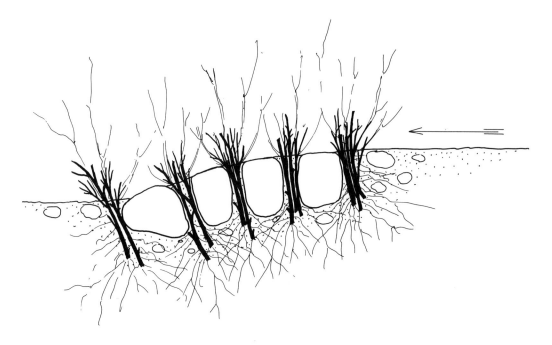

Fig. 3.9 Stone sill with live willow joint planting.

Costs
Low to moderate.

Areas of use
Brush sills are not very stable transverse structures and their use is
confined to small tributaries with intermittent flow and as a secondary
measure in torrent control to assist aggradation and bed erosion control.
They retain only relatively small amounts of fine grained bed load up to
the size of small stones.

Willow fascine sills on brush layer (Fig. 3.10)

Materials
Live fascines of 100–150 mm diameter; thicker fascines may be
constructed by including dead branches and gravel or stone; branches
capable of rooting for the brush layer.

Implementation
A simple fascine sill consists of one fascine which is buried in a shallow
trench across the bed to half or three-quarters of its diameter; the

Fig. 3.10 Live fascine sills.

effectiveness may be increased by adding a brush layer. The brush layer
is placed on the downstream side of a triangular trench dug across the
bed and partly backfilled with the excavated material; the fascine is
placed on top of the layer and the backfill completed. Higher sills may be
constructed by placing two fascines, slightly offset, on top of each other;
the effectiveness may be increased by inserting a brush layer between the
fascines.

The fascines must be secured by driving 1 m long metal pegs or stakes
at 500 mm intervals through them into the ground. Care must be taken to
anchor the fascines securely into the bank to prevent lateral undercutting, by placing their ends into narrow trenches which are reinforced
with poles driven firmly into the bank. After backfilling, the fascines may
be additionally secured by driving live willow branches into the soil on
the downstream side and spreading the protruding parts on the bank to
form a protective mat.

Timing
Dormancy period.

Effectiveness
More stable than the brush sill and not prone to scour damage.

Advantages
Simple construction.

Disadvantages
Similar to the brush sill – limited stability under adverse conditions.

Costs
Inexpensive.

Areas of use
Sensitive to moving coarse bed load, but increases siltation with fine grained sediment. Most suitable for sills in rather narrow streams of up to 5 m width and low flow velocities, and as a secondary measure to raise the bed level in torrented streams.

Wattle fence sills (Fig. 3.11)

Materials
Flexible live willow stem cuttings capable of sprouting; stakes.

Implementation
Either as a single or double sill. The fences must be buried to a minimum

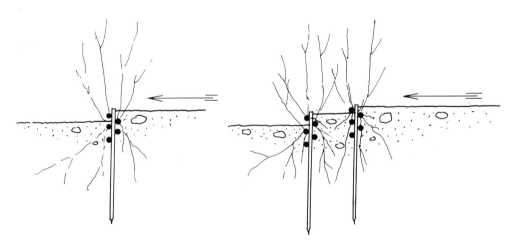

Fig. 3.11 Live wattle fence sills.

of half their height to prevent undercutting. In addition, they must be secured by driving 1 m long metal rods or stakes at 500 mm intervals into the ground on the downstream side. A strong wire mesh placed against the upstream side of the fence will protect it from abrasion and disintegration. The ends of the fence should be extended into, and a short distance up, the banks, again to prevent undercutting: the toe can be protected by the placing of a branch mat (see 'Willow fascine sills on brush layer').

It is possible to combine fence sills with brush and fascine sills. In combination with the brush sill, a fence is placed on top of the brush layer before the backfill is completed, permitting only the top of the fence to extend beyond bed level after the backfill is completed. Wattle fences may be reinforced further by placing one or several fascines on the downstream side.

Areas of use

In narrow and slow flowing streams or agricultural drains; sedimentation is increased protecting the bed level. Of limited use in torrents which do not carry a coarse bed load.

Live wire mesh gabion sills (Fig. 3.12 (left); Plates 15 and 16)

Materials

Wire mesh of 50 mm mesh size; coarse gravel; stone, binding wire; stakes and live willow branches (capable of vegetative propagation).

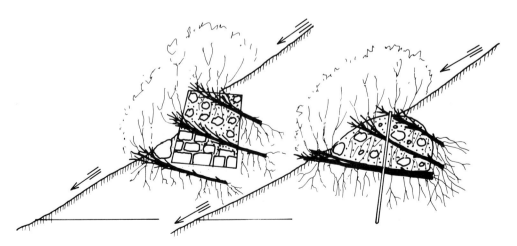

Fig. 3.12 Live wire mesh gabion and geogrid cylinder sills.

Implementation

The wire mesh is spread flat on the ground and progressively filled with stone and layers of willow branches and, if required, rooted saplings. To facilitate the proper placing of the vegetative material, the wire mesh is lifted as the fill progresses, inserting the branches through the mesh. Finally the mesh is tightly closed and secured with galvanised wire. The resultant gabion is rectangular in cross-section, fitting snugly to the ground. The branches should lie at an oblique angle in the gabion, butt ends lower than the protruding tips which face downstream. The cut ends of the branches must reach through the gabion into the soil at the upstream side to ensure that rooting can take place. As a rule, rounded and sack-like gabions are better suited for the protection of riverbanks and watercourses than the traditional rectangular shapes.

To prevent undercutting of the gabion, it may be placed on top of a brush layer. In a similar manner, the crown of the gabion may be protected by a brush layer, provided the structure is covered in earth on the upstream side. To increase the stability of the gabions, they should be partly sunk into the bed and extended into the bank on either side. If necessary, the gabions should be placed on a grid of poles to serve as foundations, the voids being filled with live branches, their tops protecting the downstream face from erosion.

Timing

Only during the dormancy period because insertion of any live branches after construction of the gabions is usually impossible.

Effectiveness

A stable sill suitable for steep and permanent watercourses carrying a heavy bed load of stone and gravel.

Advantages

Relatively simple construction using local materials, reinforced by live branches.

Disadvantages

Labour intensive construction. Provision of vegetative cover after completion almost impossible.

Costs

Lower than comparable solid structures.

Areas of use

A low barrier in torrents and mountain streams for grade control, bed erosion control and increased sedimentation. Tolerant of heavy bed loads; preferred method in the limestone areas of the Alps.

Live geotextile cylindrical sills (Fig. 3.12 (right))

Geotextile-wrapped gravel fill structures are suitable for the construction of live sills. Geotextiles and geogrids are more flexible than wire mesh and these structures may be formed in any shape to fit the terrain. Live branches are placed between the individual sack-like elements or, less desirable, live cuttings are pushed through the bags. The effectiveness and area of use are the same as those for the wire gabion sill (though durability may be less, in particular resistance to abrasion by silt, gravel and stone bed loads carried in steep fresh water courses).

Wood sills (Fig. 3.13; Plate 17)

Materials
Poles or square timber; nails; gravel and stone; willow branches and/or cuttings.

Implementation
Wood sill construction is based on a simple crib wall and should not exceed a total height of 1–1.5 m. One to five live branches per linear m are inserted between the cross pieces and the crib is progressively filled with free draining material. The branches should be sloping at an approximate angle of 15° backwards (upstream) into the natural soil, or into backfill, if necessary, to facilitate rooting.

The stability of the structure is increased if the wood frame is designed as a double crib.

Timing
During the dormancy period if the live branches are placed during construction. The crib may be constructed at any time, but vegetated only during dormancy.

Effectiveness
Stable and elastic structure. The used timber may last 25–30 years.

Advantages
Use of local materials. Quick construction. The live material at a later stage will take over the function of the decaying dead materials.

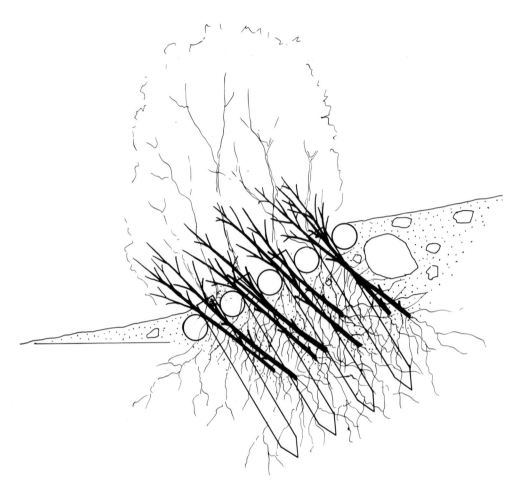

Fig. 3.13 Live timber sill.

Costs
Moderately high.

Areas of use
Suitable drop structure for gullies with fine textured bed load and narrow but steep streams, 3–5 m wide.

3.3.1.4 Live barriers

Crib walls (Fig. 3.14; Plates 17–20)

Cribs are single- or double-walled box-like structures made of timber or prefabricated concrete sections and after completion filled with free

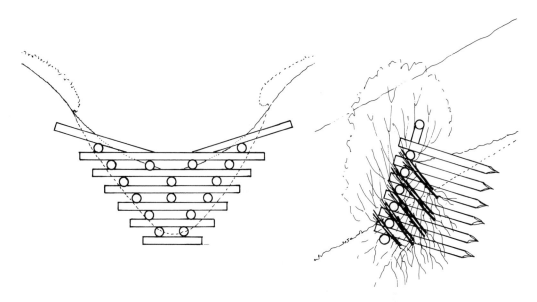

Fig. 3.14 Live timber crib wall.

draining soil. To increase stability and efficiency, freely rooting branches of deciduous shrubs and trees may be inserted into the fill material to provide a vegetative cover.

Materials
Poles or square timber of 150–200 mm diameter; precast concrete sections; nails; bolts; metal straps; sandy gravelly or stony fill material; branches of trees and shrubs, usually of willows, capable of vegetative propagation.

Implementation
On a row of poles aligned at a right-angle to the stream flow direction, similar poles are placed at predetermined intervals and at right-angles to the first line of timbers. This process is repeated until the desired structural height is reached. The poles are firmly tied together as construction proceeds. The whole structure must not be vertical but inclined against the direction of flow and securely anchored into the banks. During erection, the crib is progressively filled with soil or, if required, with stones. At the same time live root-throwing branches are placed between the poles angling downwards towards the backfill. The branches must be of sufficient length to reach the fill material or natural soil behind the crib for easy rooting, and the tips should extend beyond the face of the wall by 0.25 m.

Timing

During the dormancy period. If the construction of the crib wall has to take place during the growing season, the live materials may be inserted during the following dormancy period, albeit with great difficulty. Instead of branches, strong cuttings must then be used, but in this case the resulting plant cover is not usually as vigorous and successful as the growth achieved when large branches are incorporated whilst crib wall erection proceeds.

Effectiveness

Elimination of shear action on the bank and bed. Crib walls can withstand the forces of heavy bed loads. The use of treated timber increases the life of the crib to 50–60 years. Established vegetation should maintain the level bed sections, at least in intermittent flowing streams, even if the crib timber has deteriorated over time. The crib wall by then will have changed into small, vegetatively reinforced soil drop structures.

If more durable concrete elements are used for the construction of the wall, the established vegetation will create suitable habitats for flora and fauna (though alkalinity and drought scorching can occur).

Advantages

Construction material locally available (in the case of timber cribs in forested areas). Construction relatively easy and quick.

Costs

Cheaper than comparable solid structures. Live barriers are drop structures designed to reduce steep gradients of watercourses through aggradation, but they cannot serve as bed load barriers. Maximum construction height up to 5 m.

Modified crib walls were used to great effect up to the 1940s for the construction of diversion weirs and collection points serving saw mills and small power stations and for the transport of timber logs.

Wire gabion barriers (Plate 21)

The wire gabion barriers developed in the Italian Dolomites were used for the construction of crossing points of torrents and mountain streams with intermittent flow by placing as many gabions as needed to achieve the desired height. The layers of gabions were placed offset and the whole structure built at an angle for greater stability. Further development work led to the construction of barriers and weirs. The basal area of the rectangular gabions used measured 1–1.5 × 3–5 m. The total height of such structures should not exceed 3 m. Special attention must be paid to the foundations and lateral protection of such structures.

Materials
Zinc coated wire mesh of 50 mm mesh size; rubble and gravel; large stones; binding wire; vegetatively propagating willow branches and stakes; rooted saplings.

Implementation
If only a single row of gabions is to be placed, the live branches are inserted as outlined in Section 3.3.1.3 'Live wire mesh gabion sills'. If several layers of gabions are required, the branches are placed between the gabions. The tips of the branches should extend by a length of approximately 250 mm on the downstream side, and be firmly buried in the soil on the upstream side.

Timing
Dormancy period (low water period), because placing of the vegetative material at a later stage would be of limited success and incur high costs.

Effectiveness
Solid and robust structure with some flexibility for the retention of heavy and coarse bed loads.

Advantages
Stone, gravel and rubble usually available on site. It is possible to repair works if necessary.

Disadvantages
Gabion structures are of lesser importance in alpine areas with siliceous rocks, as these can break down over time and the gabions collapse.

Costs
Less than comparable solid structures.

Areas of use
Transverse drop structures to achieve level bed sections in streams and torrents. Construction of permanent crossing points.

Dry stone wall barriers (Plate 22)

Dry stone walls using natural stone are built at a slight angle, never vertical. Foundations and lateral keying follow the principles laid down for weir construction. Walls of a height exceeding 5 m are nowadays constructed without difficulties.

Materials
Stone and quarry stone; sandy-gravelly soil; rooting branches and/or saplings; container plants; turf from natural grassland; grass seed.

Implementation
The live plant material is placed during construction between the stones, observing the guidelines laid down above for wire gabion barriers. The joints between the stones are filled with the sand–gravel mix. Container plants and natural turf may be placed in a similar manner.

Timing
Only during the dormancy period if rooting live plant materials and open rooted saplings are used. Container plants and natural turf may be placed during the vegetative season.

Effectiveness
Solid and massive structure to retain heavy bed loads.

Advantages
Of use in all areas where large quantities of natural stone are available.

Costs
Less than comparable solid structures.

Areas of use
Dry stone walls are counted amongst the most stable transverse structures. They withstand heavy stress and are capable of retaining large soil masses and heavy bed loads, and are most suited for the control of large steep gullies, and serve as primary transverse structures to achieve level bed sections in torrent control work.

3.3.2 Deflectors

Deflectors are designed to protect the riverbank by diverting stream flow and dissipating its energy. As a rule they are used in rivers with a bed width of more than 10 m.

3.3.2.1 *Groynes, dykes and spurs* (Fig. 3.15; Plates 23–31)

Groynes are small dykes reaching from the bank into the river to create a uniform stream bed. The outer end or the 'head' of the groyne has to absorb the energy of the flowing water, and the areas between the groynes serve as sediment basins. Modifications of the flow may be achieved by extending or shortening the length of the groyne. The angle of the groyne in regard to the direction of flow may vary by a certain amount from the right-angle, either upstream (inclined) or downstream (declined); the latter, downstream-facing groyne is rarely used, because when flooded they create conditions similar to those of an overtopped weir. The flow downstream of the groyne is directed towards the bank, possibly resulting in erosion damage. The angle of inclined groynes, i.e. those pointing upstream, amounts to 70–85° on flooding; the direction of the flow is then towards the centre of the stream, away from the sensitive bank. If groynes are constructed on both banks, the heads should be opposite each other, otherwise the flow is directed into the still water areas between the groynes (Lange and Lecher, 1989).

The spacing of the groynes along the banks should not be wider than the distance between two groyne heads on opposite banks. When the crowns of the dykes are not submerged during normal discharge periods, water flows in a circular pattern between two adjacent groynes. The confined flow between two groyne heads spreads downstream and hits the next groyne, causing the circular motion in the basin between the groynes and at the same time an appreciable reduction in velocity, which leads to increased sedimentation. Due to the circular water movement, sediment distribution is rather uniform in the area between the groynes. The angle of flow deflection from the groyne head into the basin amounts to 6° (Lange and Lecher, 1989). Therefore short groynes should be closely spaced; if the distance between groynes becomes too short, bank protection by longitudinal structures is more economical and more effective.

The height of the groynes is calculated to coincide with the mean average flow level. Higher flow levels will submerge the groyne, which necessitates adequate protection for crown and side slopes. The upstream slope should be between 1(V):2(H) and 1(V):3(H), the downstream slope 1(V):3(H) and 1(V):4(H). The gradient of the groyne head facing the centre of the river varies with the type of stream and should be 1(V):4(H) to 1(V):10(H).

Several methods are employed in the construction of groynes and the choice of building materials is extensive, including metal sheet piles, concrete pile poles or timber logs, rip-rap, wire gabions and many others,

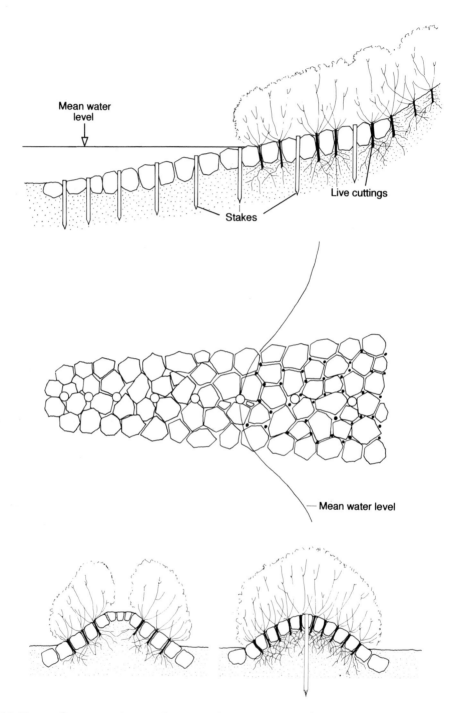

Fig. 3.15 Types of groynes. *Above and centre*: submerged groyne of rip-rap rock and live stakes. The joints above mean water level are filled with soil and planted with willow cuttings. *Below left*: trapezoidal groyne. *Below right*: rounded groyne.

which may be used on their own or in combination, if necessary reinforced by bioengineering techniques.

Originally, groynes were used to confine widely meandering rivers in order to obtain agricultural land not subject to flooding. Very long dykes were constructed using, as a rule, local materials such as stone and gravel, wood poles and branches obtained from woodland and coppices in the flood plain. Over the last 100 years, groynes were mainly constructed for the protection of riverbanks and the repair of slump failures.

The availability of sufficient land is a precondition for the construction of ecologically effective groynes and dykes as a measure in the management of rivers and watercourses. At the present time, groynes must be considered as a compromise between a natural river corridor with a meandering watercourse which, due to lack of space, is not an acceptable proposition, and a regulated and straightened or canalised riverbed.

Materials
Rocks; boulders; quarry stone; gravel; poles; wire gabions; branches of rooting bushes and trees for vegetative erosion control.

Implementation
This depends largely upon the quantity and quality of locally available materials, with two types of structures dominating the field: *rock-filled log crib* and *blockwork*. The former are wooden structures made of logs or square timber similar to double cribs filled with quarry stone to provide the necessary stability. Tops of groynes that are not permanently submerged may be reinforced by the insertion of root-throwing branches of suitable tree species after mixing some fine gravelly soil with the stone to provide a rooting medium. To prevent the fine material from being washed away, it could be contained in sand bags or similar textile containers. Rock or blockstone groynes consist of various stone structures into which long root-throwing branches are inserted above the mean water level during construction in the dormancy period. Willow cuttings are inserted in the joints of the crown. For hydraulic and aesthetic reasons, the contours of the groyne should be rounded, not angular. Timber piles are usually incorporated in groynes – they are pre-driven before stone blocks are placed.

Both types of groyne should decrease in height toward the stream centre. The downstream side of the groyne should slope at a steady gradient if necessary to bed level to ensure that part of the structure is always in contact with flowing water.

Timing
During low flow periods, i.e. in the dormancy period.

Effectiveness
Groynes may replace longitudinal bank protection works, but must be of reasonable length to keep costs down. Their main function is to direct flow from the banks, and the established vegetative cover will reduce flow velocities and dissipate energy. Groynes will have a certain regulating effect without confining the flow into a single channel as is the case with longitudinal dykes. The cross currents and circulating water masses in the areas between two adjacent groynes cause alternating cycles of bed erosion and deposition and create biotic conditions and diversity that are not quite of the same order as those obtained in natural and undisturbed river courses, but are superior to those prevalent in regulated rivers. In addition, they provide easy and safe access to the water for recreational purposes.

Advantages
Groynes are flexible structures which may be modified, lengthened or shortened with modest capital expenditure. They have a long life and low maintenance and repair costs. The riverbanks need a minimum of protection which can be achieved by vegetative means.

Disadvantages
During high flood periods excessive cross currents may cause erosive eddies, and undercutting of the groyne heads may necessitate periodic higher maintenance costs. Groynes take up more space than longitudinal bank protection works.

Costs
Lower than continuous solid bank protection measures.

Areas of use
Ecological and technically sound riverbank protection measures for watercourses wider than 10 m. Maintenance of biodiversity on land and in the water.

3.3.2.2 *Live brush sills* (Figs. 3.16 and 3.17)

Live brush sills are mainly used in conjunction with reinforced brush barriers or brush traverses (see Section 2.3). As soon as the scour has

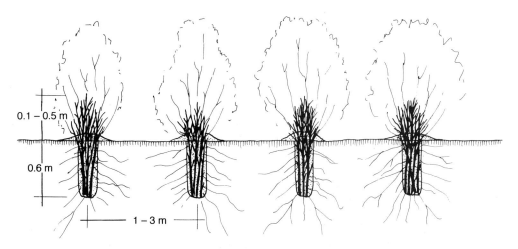

0.1 – 0.5 m

0.6 m

1 – 3 m

Fig. 3.16 Live brush sills for promoting siltation.

Fig. 3.17 One-year-old willow brush sills on top of riverbanks behind stone block revetment.

been filled with sediment to the level of the mean summer water mark, several rows of live brush lines are planted between the brush barriers using willow cuttings. In many cases, sedimentation decreases with increasing distance between the failed bank and the river centre; the cuttings should therefore be planted initially in close proximity to the bank and the lines extended toward the centre line of the stream at a later stage, keeping pace with the progressive rate of sedimentation.

Materials
Branches, cuttings and stakes of root-throwing woody plants.

Implementation
The live branches and/or cuttings are planted in rows 1 m apart and 100–150 mm thick; closer spacings may be chosen if required (Fig. 3.16). The branches protrude from 100–200 mm up to 500 mm above soil level. The rows are angled at 30° against the direction of flow and the cuttings are inclined downstream.

Timing
During the dormancy period. If the branches are stored in flowing water, during the early and late growing period.

Effectiveness
Reduces the tractive forces, encouraging sedimentation. The roots provide protection against scour and erosion.

Advantages
Simple and quick method.

Disadvantages
Limited to flat alluvial areas. Not very stable on its own.

Costs
Very inexpensive method.

Areas of use
On their own, brush sills are not as effective as brush traverses, nor are they as robust. They supplement, however, the effect of the brush traverse and encourage sedimentation during the critical period when the siltation in the scour area has progressed to, or is reaching, the level of mean summer flow. They are used mainly in conjunction with traverses to assist in the final silting up of extensive scour areas.

3.3.2.3 *Brush traverses (Keller, 1937)* (Figs. 3.18–3.20; Plate 32)

Brush traverses are a type of groyne and are a much larger variant of the live brush sill.

Materials
Quarry stone; crushed stone; cylindrical wire baskets; fascines; stakes

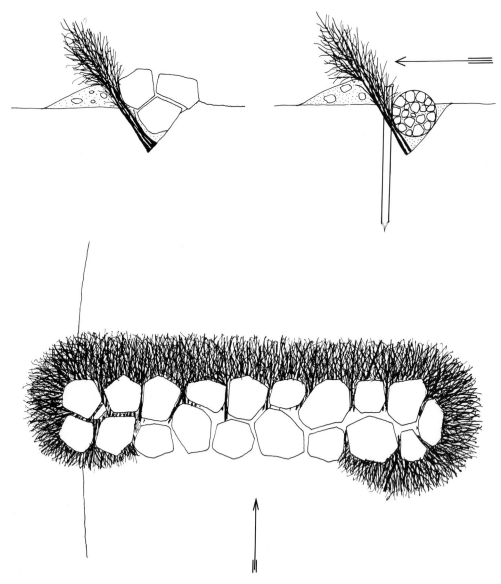

Fig. 3.18 Construction details of brush traverses.

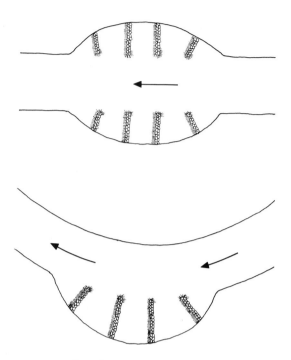

Fig. 3.19 Arrangements of brush traverses along watercourses.

and root-throwing branches mainly of willows and poplars, 1–1.5 m long.

Implementation
Approximately 0.5 m deep, asymmetrical (triangular) trenches, with the flatter slope facing downstream, are dug and the 1–1.5 m long branches are placed close to each other facing downstream in the trench. The branches may be packed very closely, resulting in an intertwined mat. Any gaps must be carefully avoided because the accelerated flow through the gap may cause serious erosion which may ultimately lead to the destruction of the traverse. To facilitate root formation, the trenches may be partly backfilled with topsoil. Large stones or pegged down cylindrical wire gabions are placed on top of the trench to firmly secure the branches. The traverse should be keyed deeply into the bank, filling the trench with large stones. In addition, extra branches should be pushed through the gabions or stone layer in close proximity to the bank to form a protective fan. The river end of the traverse should be secured by extra large stones, reinforced with live branches placed at a right-angle to the direction of flow in order to dissipate the water's energy, taking care not to leave any gaps in the branch layers.

Fig. 3.20 Brush traverses after completion of construction.

The sedimentation and silting of a seriously eroded bank through scour is achieved by a series of traverses. The first traverse at the upstream side of the damaged bank is placed to form an acute angle with the projected bank alignment to divert the flow towards the river's centre; the adjacent traverses are at right-angles, and the last one at an obtuse angle to the bank to encourage uniform sedimentation (Fig. 3.19). The distance between the traverses is kept to the length, or one-and-a-half times the length, of the traverse.

Timing
During the dormancy period, which usually coincides with minimum river flow.

Effectiveness
The elastic branches of the traverse dissipate the energy of the flowing water and reduce the tractive force, resulting in accelerated sedimentation. The rate of sedimentation is further increased by the number of traverses; at the same time, the base load is subject to a certain sorting, with the fine fractions accumulating at the downstream end of the

protection works. The established vegetative cover will ultimately lead to the complete siltation of the eroded area.

Advantages
Simple construction. Fully effective after a short time. Very economical method. Resistant to tractive forces of up to 150 N/sq m. Simple maintenance and, if necessary, repair of traverses, which may be extended if required.

Disadvantages
Not suitable for torrents with heavy bed loads. Construction only possible during the dormancy period.

Costs
One of the most economical methods to repair stream bank erosion by inducing sedimentation.

Areas of use
General stream bank protection in the area between low water and mean high flood level and repair of scour areas in rivers with moderate bed loads. Regulating rate and degree of sedimentation caused by natural or artificial (boating, water sport) currents along flat lake shores.

For very fast flowing rivers, the brush traverse may be combined with the brush grid method for greater effectiveness (Prückner, 1965). The brush grid is constructed at the upstream edge of the scour area to divert and dissipate the energy of the most aggressive flow and, at the same time, initiating sedimentation. The brush traverses constructed downstream in more steady water will increase the rate of siltation. If the planned alignment of the bank to be reconstructed is subject to continuously strong erosive forces, as is the case in the outer curve of a bend, a series of strong poles at 2–3 m centres are hammered into the bed. Horizontal laths or thin poles are then fixed to the stakes every 200–300 mm. The lowest lath should be at the low water level, and the highest just above the level of the average summer flow. It is the main function of this structure to break the force of the flow entering the scour area and deflect the main stream to the centre of the planned riverbed, at the same time excluding floating debris, etc. from the reconstruction zone.

3.3.2.4 Brush grids (Prückner, 1965) (Figs 3.21–3.23; Plates 33 and 34)

Materials
Poles; quarry stone; dead branches and small trees for the lowest layer;
live root-throwing/shooting branches for the upper layer.

Implementation
First, a line of strong poles is hammered into the bed along the line of the
projected bank alignment; the distance between the poles should be kept

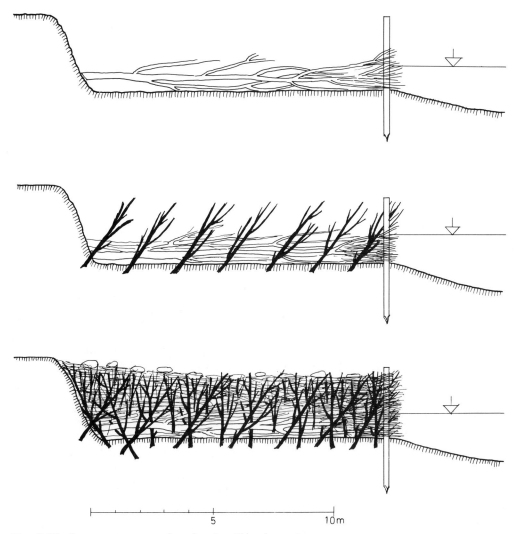

Fig. 3.21 Construction stages for a brush grid bank repair.

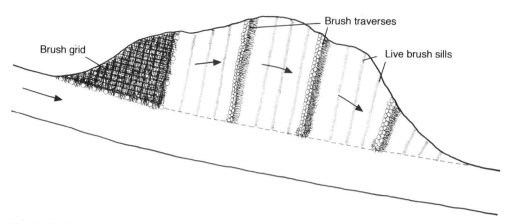

Fig. 3.22 General arrangement of upstream brush grid, brush traverses and live brush sills for repair of large-scale bank erosion.

Fig. 3.23 New brush grid repair of eroded bank before natural silt infilling.

at 2–3 m. The dead branches and small trees are placed close to each other at right-angles to the projected bank alignment in the area between the line of poles and the eroded bank. The tips of the branches and trees should protrude for a distance of 0.5–0.75 m beyond the line of poles. The thickness of the dead branch layer depends upon the depth of water in the eroded area.

Next, live cuttings arranged in parallel lines are pushed deeply into the riverbed at an oblique angle facing downstream. To stabilise the branch layer, the cuttings near the line of piles demarcating the proposed new shoreline should be embedded particularly deep into the riverbed. The live branches should then be secured by placing large stones or boulders at 1–1.5 m intervals on top of the line. Further lines of live branches arranged at right-angles to the first line are established in a similar manner, with the tips of the branches again slanting downstream, and weighed down with large stones. The entire grid-like structure is then lightly covered with soil. The surface of the completed branch layers should be approximately 1 m above the mean low water level.

If a shortage of live and dead branches prevents the establishment of a closed mat, the structure may be arranged in discreet lines forming a grid.

As a final measure, the face of the failed bank should be covered with a dense layer of willow branches resembling a brush mat to prevent further erosion or slumping.

It is usually not necessary to cover the total surface of large bank failures with brush grids. Only the top part needs to be secured with the brush grid; the downstream section may be protected by brush traverses or simple brush sills.

Timing
Dormancy or early and late growing period during low water.

Effectiveness
The very stable brush grid is mainly used for the rehabilitation of severe erosion damage extending over a large area. The method is suitable for the establishment of new bank alignments. It is resistant to high floods and causes rapid sedimentation by reducing the tractive forces.

Advantages
Immediately effective. Very stable structure.

Disadvantages
Very labour intensive.

Costs
Average amounting to 25–30% of corresponding solid structures.

Areas of use
Rehabilitation of large bank failures and eroded areas, particularly on fast flowing rivers and streams subject to large fluctuations in water levels. Appropriate for water depths of up to 3 m.

(II) Longitudinal structures

Longitudinal structures for the protection of stream banks are required where lack of space precludes the adoption of other measures. Entirely mechanical or 'solid' structures do not always fit well into the landscape and are perceived as alien. This perception of something unnatural can be largely modified by the combination of mechanical measures with bioengineering techniques to enhance their appearance. The following sections give an outline of such techniques.

3.3.3 Reed plantings (Figs 3.24–3.26; Plates 35–39)

Reed plantings are particularly suited to the protection of the shorelines of lakes and dams, canals and other slow moving water bodies against wave action. Experience has shown that the reed grasses (*Phragmites*), ribbon grass, bulrush (*Schoenoplectus lacustris*) and several sedges (*Carex*) are eminently suitable for the purpose (see Section 1.3). Reeds may be planted as single stem, rhizome or in clumps or reed rolls. The purifying action of the various reeds on polluted waters has been investigated in various countries (Seidel, 1965; 1971).

3.3.3.1 *Reed single stem or culm planting* (Fig. 3.24 a–c; Plates 35 and 36)

Reed grass (*Phragmites communis*) is the most commonly used, although others such as ribbon grass, are equally suitable.

Materials
Hand-placed quarry stone or rip-rap; stems and culms which form adventitious roots at the nodes.

Implementation
Young and vigorous plants, 800–1200 mm tall with a maximum of five

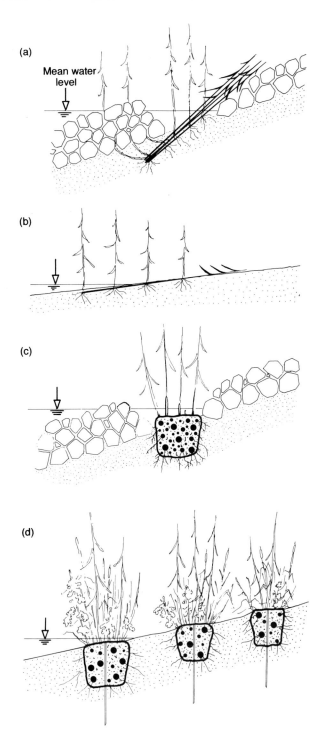

Fig. 3.24 Reed transplants along shorelines. *(a)* reed stem planting between rip-rap and *(b)* on bare soil; *(c) and (d)* reed rootball planting combined with rip-rap *(c)* and on bare soil *(d)*.

Fig. 3.25 Reed rhizome planting. *Above*: after one year. *Below*: after two years. (Courtesy Nawrot, University of Illinois, USA)

leaves are dug up with their roots from the edges of existing reed beds. To prevent accidental damage to the sensitive plants, they are bundled and wrapped in hessian or similar material for transport to the construction site. The stems are planted in rows, three to five culms together per planting station every 0.25–0.5 m. It is of advantage to plant the stems at an angle so that they almost touch the ground. This prevents the culms from being damaged by wind or wave action. The stems are planted some 100–150 mm below the mean summer water level, preferably under the water surface. The reed may be planted straight into the soil, or be inserted in the joints of hand placed stone. If rip-rap is to be placed on the bank, the plants should be placed into narrow trenches which are left free of stones but backfilled with soil and fine gravel; for best results, the stone work on the water side should be slightly raised to protect the growing plants (Fig. 3.24).

Timing
The planting period is relatively short, lasting from the beginning of May to mid-June (se Fig. 2.3).

Effectiveness
The culms and leaves of the plants dissipate the energy of flowing water or wave action and protect the bank or shoreline from erosion, and the roots effectively bind the soil. Accelerated siltation of the inner bank of a winding river course is often a disadvantage. The full effect of reed plantings is only realised after the second or third year of establishment.

Advantages
Simple and economical construction.

Disadvantages
Only short time space for establishment. Protective action not immediate. Will not tolerate shade.

Areas of use
Establishment of reed beds on the banks of slow moving watercourses or stagnant water, flat lake shores, preferably on silty, sandy locations in full sun. Establishment on shores or banks subject to strong wave action may be difficult. Plantings should always proceed from the wet shoreline toward the dry ground.

3.3.3.2 *Reed clump plantings* (Fig. 3.24d; Plate 37)

Materials
Hand-placed stone or rip-rap; clumps; rhizomes of local reeds, mainly
reed grass. Sods of approximately 300 mm cubic dimension or larger if
possible are cut from existing reed beds, preferably from the land side.
Cutting is best done during the dormancy period or during the growing
period for immediate use; the optimum planting time is before the new
shoots appear in spring, that is, between the beginning of March and
mid-April (Fig. 2.3).

Implementation
Care must be taken not to damage the young shoots during transport to
the construction site. The clumps are planted in close contact with each
other in trenches if sufficient sods are available, or individual clumps are
planted in holes some 500 mm apart. Reed grasses such as reed
(*Phragmites*), rush (*Juncus*), sweet grass (*Glyceria* spp.) and bulrush
(*Schoenoplectus lacustris*) sods are planted flush with the soil surface just
below the level of the average summer water; the culms of sedges (*Carex*
spp.) and reed canary grass (*Phalaris arundinacea*), just above this level.
Sods obtained from lake shores must be wrapped in hessian, jute or other
biodegradable material, taking care to let the stems protrude above the
packing material.
   The clumps are planted into the revetment or into the unprotected
stream bank. If planted in a revetment, the clumps must be in contact
with the soil underneath and should only be covered lightly with stone to
facilitate rooting and shooting.

Timing
During the dormancy period; for best results before the new shoots
appear in spring. If carefully transported, the clumps may be planted up
to the end of April.

Effectiveness
Reed clumps become effective much sooner than stem plantings on
account of the larger root reserves. It takes two to three years for the
plantings to become fully effective. After this time, there is no difference
in the vigour of the plants established by either method.

Disadvantages
Short planting period. High transport costs due to the heavy weight of
the sods. Not tolerant of shade.

Areas of use
On the banks of fast flowing or slow flowing rivers, canals, lakes, dams, etc., but not on banks or shores subject to heavy wave action. On the cut bank of non-navigable rivers, but not on the point bar because accelerated siltation may divert the main flow toward the vulnerable cut back. Navigable canals and watercourses should have both banks protected with reed zones to protect the banks from wave action caused by the passing boats.

3.3.3.3 *Reed rolls* (Fig. 3.26; Plates 38 and 39)

Materials
Clumps of various sizes of reed grass but also of bulrush (*Schoenoplectus lacustris*), floating sweet grass (*Glyceria fluitans*), sweet flag (*Acorus calamus*), yellow flag (*Iris pseudacorus*), meadow sweet (*Spiraea ulmaria*), calla (*Calla palustris*) and sedges (*Carex* spp.); poles; wire netting; live geotextile cylindrical sills; gravel or crushed stone; boards; planks.

Implementation
1–1.5 m long wooden poles are hammered into the soil at 1–1.5 m intervals along the line coinciding with the mean summer water level, so that approximately 300 mm show above the water level. Parallel to and behind the line of stakes, a trench 0.4–0.5 m deep and of similar width is dug. If there is any sign of the trench walls slumping, they should be supported by wooden planks. Wire netting is placed into the ditches and the triangular voids at the bottom of the trench between wall and net filled through the mesh with soil. The netting is filled with a mix of coarse gravel (20–60 mm diameter) and crushed stone (60–120 mm diameter) and the vegetative bits and pieces left over from trimming the reed clumps. The top third of the netting is filled with reed clumps and stitched up with wire, resulting in a cylindrical body. The planks supporting the trench are then removed, the gaps filled with soil and the stakes or poles hammered further into the ground to just below the apex of the reed roll. After completion of the works, the roll should protrude for 50–100 mm above water level. The rolls are of approximately 300–400 mm diameter. Extensive bank failures with near vertical walls below the water level are to receive further protection by placing a buried fascine or crushed stone between the reed roll and the water's edge.

Timing
During the dormancy period; best in spring before the reeds start throwing roots.

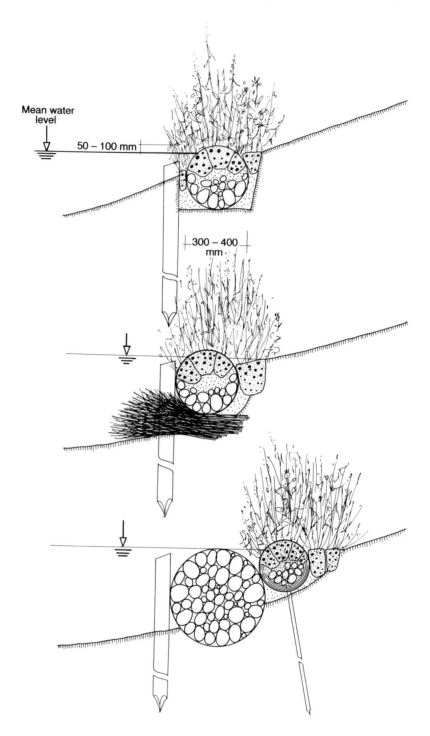

Fig. 3.26 Forms of reed roll construction. *From top to bottom*: without protection against undercutting; protected by a brush layer; protected by a cylindrical gabion.

Effectiveness
Technically the most stable of the reed structures. Affords protection
against bank erosion.

Advantages
Immediately effective.

Disadvantages
Limited construction time. Labour intensive method requiring large
amounts of materials and therefore the most expensive method amongst
the reed structures.

Costs
4–6 work hours per linear m.

Areas of use
Protection of stream banks and small rivers with fairly constant flow
rates and fine grained bed load. Protection of newly established reed
plantings on lake shores prone to erosion, drainage channels, etc.

3.3.4 Structures with woody plants capable of vegetative propagation
(Figs 3.27 and 3.31; Plates 40–53)

3.3.4.1 Stone revetments reinforced with cuttings (Fig. 3.27 and Plates 40–43)

Materials
Large irregular shaped sone; cuttings 0.5–1 m long depending upon size
and area of the stone work; backfill material.

Implementation
The joints between the stones should be as narrow as possible; however,
above the average summer water line, larger gaps should be left every
0.3–0.5 m to accept the cuttings. Planting holes may be made, if necessary, using a steel rod. The cuttings should not protrude more than 50–
100 mm and should, if necessary, be trimmed to that height. Any voids
and gaps must be filled with soil.

Timing
Planting of cuttings only during the dormancy period, if necessary, in a
separate operation after completion of the stone pitching.

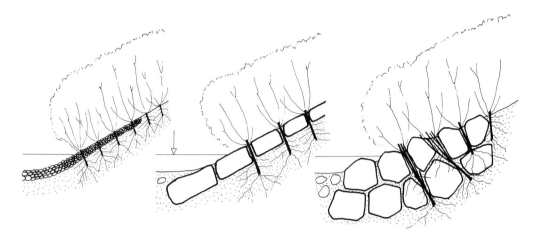

Fig. 3.27 Use of shrub cuttings for riverbank protection. *From left to right*: stone armour toe layer cuttings; stone revetment with interstitial or joint planting of cuttings; rip-rap with branch layers.

Effectiveness

Root development takes place in a short space of time. Roots and shoots provide a quasi-reinforcement to the stone layer; this enhances the effectiveness of the measure as the erosive action of the water is weakened. For this reason, vegetatively supplemented mechanical structures such as stone pitching are adequate even if smaller stones are used. The often expressed reservation that stone pitching is weakened by the planting of cuttings in the joints is largely baseless. On the contrary, the increasing mass of stems and branches and the intensive root system consolidate the structure. After several years, the stone revetments cannot be seen any more, leaf litter and accumulated organic matter form a humus layer, and a closed vegetative cover results from the planted, widely spaced cuttings, providing a suitable habitat for animal life.

In a dry climate, cuttings take better when planted in joints than on open slopes and watercourse banks. Flood water may lead to a loss of soil from the joints, causing a certain amount of failures amongst the cuttings. A failure rate of 30–50% must be expected. Soil loss from the joints below the mean water line can be avoided by incorporating a geotextile filter during construction.

Advantages

Very effective and simple conservation method using local plant materials.

Disadvantages
Full ecological and vegetative cumulative effect only developed after several years.

Costs
For the collection and planting of the cuttings, together with all ancillary work, 2–5 work hours per sq m.

Areas of use
Massive and stable protection of banks of fast flowing rivers and streams with active wave action and heavy bed loads. Protection of lake shores where high winds and/or boat traffic cause increased wave action.

Container or open rooted plants may be used for joint planting, if the spaces between the stone revetments are filled with topsoil. Planting is best undertaken during spring and autumn, rarely throughout the season. The use of several species of rooted plants will result in a more diverse vegetation cover as compared with the planting of cuttings alone, which usually forms a uniform stand of willows.

3.3.4.2 *Branch layers* (Fig. 3.27 (right) and 3.28)

Materials
Natural stone; live shooting branches of shrub willows of 1–1.5 m length.

Implementation
The branches are inserted during construction in crushed stone or riprap so that the butt ends are imbedded in the soil behind the cover to a depth of 0.3–0.5 m. The branch tips should protrude for approximately 0.5 m from the stone layer and face at an oblique angle downstream. The lowest branches are below the level of the mean summer flow and well wedged amongst the stone to prevent them from being washed out.

Timing
During the dormancy period, if the branches are stored in flowing water, up to the early growing period.

Effectiveness
Very stable structure. Provides immediate protection. Effects and results of combined use of live and solid construction materials as those outlined in Section 3.3.4.

Fig. 3.28 Rip-rap with brush layer after completion.

Advantages
Use of locally available plant materials. Ecologically and technically efficient method.

Disadvantages
Branches cannot be inserted after completion of the dumped stone cover when only limited use of cuttings can be made.

Costs
Lower than stone pitching with live cuttings.

Areas of use
Protection of the banks of fast flowing rivers with large fluctuations of the water level and heavy bed loads, such as mountain streams, etc. Repair of bank erosion and bank failures. Protection of lake shores subject to strong wave action.

3.3.4.3 *Live fascines* (Fig. 3.29; Plates 44 and 45)

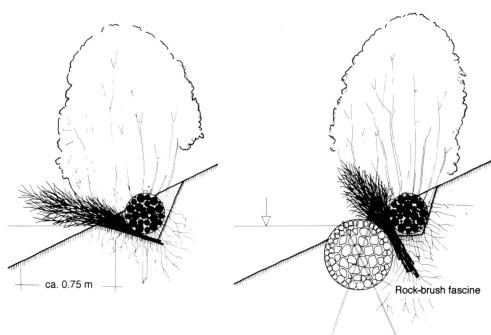

ca. 0.75 m

Rock-brush fascine

Fig. 3.29 Construction of live fascines for toe protection of steep shorelines.

Materials

Live branches and long stem cuttings of brush and tree willows, 1–3 m long; dead branches; binding wire or straps.

Implementation

Fascines are usually made up on site with the help of a contrivance resembling a saw horse. The mix of branches and switches is placed between the cross timbers and tied with flexible willow branches, wire, plastic straps or similar material every 0.3–0.5 m to form cylindrical bundles of 2–3 m length. Alternatively the fascines may be wrapped in wire mesh or geotextile or geogrid reinforced structures, at somewhat higher cost. Single fascines may be joined end to end to produce long cylindrical bodies. On site, the prepared fascines of 150–300 mm diameter are placed into trenches following approximately the line of the average summer water level, buried up to one-half to two-thirds of their diameter in the soil, sometimes below the water level. Live or dead stakes approximately 0.75 m long are driven through the fascines at 1 m intervals, anchoring them firmly to the ground. If sufficiently large quantities of live material are not available, dead branches and parts of non-vegetatively propagating bushes may be included in the fascine; such materials are then placed in the centre of the fascine. To protect the fascine from wave action or undercutting, it may be placed on a brush layer with the tips of the branches protruding for 0.5–0.75 m and pointing toward the bed centre. If necessary, higher riverbanks may be secured by placing several fascines on top of each other to form a 'fascine wall'.

Timing
Dormancy period.

Effectiveness
Immediate protection against wave attack and fast flowing water. The growing elastic branches bend easily and the rising water level or wave action will cause them to form a protective cover on the bank. The root system will consolidate the bank by binding the soil. The growing bushes and trees provide suitable habitats for animals and aquatic life.

Advantages
Quick and simple construction.

Disadvantages
Construction only during the dormancy period.

Areas of use
For the protection of riverbanks and shorelines on their own or in combination with other methods. To reinforce or stabilise other bioengineering techniques (see Section 3.3.1.3).

So-called 'sink fascines' contain a centre of gravel or rushed stone wrapped into branches of live willows, etc., tightly tied with steel wire, chains or similar materials. The heavy rolls are constructed on site and lowered to the toe of the bank, which is, if necessary, reinforced with a line of poles driven into the bed to fix the fascines in place. The top fascine should reach the level of the average depth of the summer water level or slightly beyond it. Effectiveness, advantages and areas of use correspond to those of the live fascine. Their disadvantages are the relatively high cost and expertise required for construction to achieve optimum results.

3.3.4.4 Branch packing (Fig. 3.30; Plate 46)

Branch packings are constructed in many variants and are known under many names which mean different things to different people. In some instances, the same name may mean different structures; in others, various names are used for the same method. However, all these methods are based on the construction of one or several layers of live willow branches for the repair of riverbank failures, and will be treated under the collective name of 'branch packing'.

Materials
Live branches of indigenous willows capable of vegetative propagation; branches of other non-vegetatively propagating bushes and trees; stakes; poles.

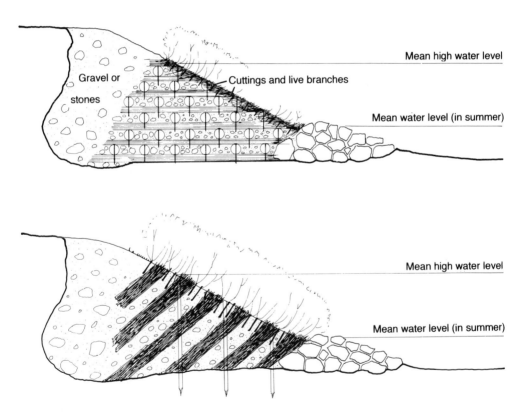

Fig. 3.30 Sections through alternative layouts for branch packing installations. *Above*: classical construction method with dead fascines arranged in a criss-cross manner in horizontal layers with live cuttings and branches installed over the surface. *Below*: simple branch packing using alternate layers of live and dead branches in inclined layers.

Implementation

Small and rather shallow bank failures are repaired by hammering 600–800 mm long stakes into the bank at approximately 1 m spacing. A dense and thick layer of live branches is packed between the stakes, with the butt ends at bed level. The stakes are braced together with wire and driven further into the ground, thereby compacting the branch packing to a thickness of some 100 mm. The branch layer is then lightly covered with topsoil, filling all voids and spaces between the branches. Larger eroded areas and bank failures are treated in a similar manner, placing one or several branch layers more or less horizontally between the line of stakes. Several method of placing the branches are possible:

❑ The branches are placed at random without direction.
❑ The branches are placed facing the same direction, preferably at a right-angle to the shoreline, the butt ends towards the land side.
❑ Different layers are placed at right-angles to each other, with the bottom layer placed as above.
❑ Alternate branch and fascine layers are placed at right-angles to each other

The individual layers must be covered with topsoil, taking care not to leave any empty spaces to avoid collapse or subsidence within the structure. Additional cover with large stones is of advantage.

When placing the branch packing, the level and slope of the completed structure must be the same as that of the adjacent undamaged parts of the bank upstream and downstream of the failure. After the placing of the branch layers, the stakes or poles must be braced together with strong wire and hammered further into the ground until there is no springiness left. If there are sufficient quantities of live and vegetatively propagating willow branches, non-vegetatively propagating branches may be blended with the live ones. In multi-layer packs, the lowest below the level of the mean summer water level should be made up of shooting/non-shooting or dead branches. In mixed single layers, at least 25% of all branches should be capable of vegetative propagation.

Timing

During the growing period.

Effectiveness

Immediately effective in protecting repaired banks. Velocity and tractive force of the water is diminished by the new stem growth, preventing further erosion.

Advantages
Simple and quick method using local materials.

Disadvantages
Large quantities of materials required.

Costs
If live branches are available in close proximity to the site, costs are low.

Areas of use
Branch packings are massive and stable structures which not only give immediate protection but also increase in efficiency with the passing of time. Very suitable method for the repair of bank failures on slow or fast flowing rivers or impounded waters. Which of the various methods should be selected depends on the site characteristics, flow velocity, depth of water and slope of the bank.

3.3.4.5 Live river training structures

Live river training structures are required to contain fast flowing rivers and streams carrying large bed loads of coarse materials. They are very effective where other more conventional methods are not feasible, e.g. within built-up areas. Flow velocity, tractive forces and the amount and nature of the bed load determine the type of structure to be chosen.

Live brush training structures (Prückner, 1965)

Materials
Quarry stone; live branches of indigenous brush willows.

Implementation
In its most simple form, the structure consists of live willow branches, cut to length to fit the height of the failed bank section, placed in close proximity against the surface of the failed bank, with the butt ends firmly pushed into the bed. The base of this wall of branches is reinforced by pushing short willow branches obliquely into the bank at a right-angle to the direction of flow to protect the toe against scour. Finally, the bases of all the branches are anchored by a layer of large stones. If greater stability is required, an asymmetrical triangular trench of 0.3–0.5 m depth is excavated at the base of the failed bank section, the side with the flatter angle facing the river, before the branches are placed against the

face of the failed bank, with the ends at the bottom of the trench. The flat slope of the trench is covered by a criss-cross layer of willow branches. The trench is then backfilled covering the bases of both branch layers, and additionally secured by placing a layer of large stones or cylindrical wire gabions on top. This structure combines the elements of a brush mat with the standard brush traverse arranged parallel to the bank.

Timing
During the dormancy period, if the branches are stored in flowing water, the construction period may be extended into the early growing season.

Effectiveness
The two branch layers effectively reduce flow velocity and tractive force by dissipating energy during flood periods. Loosened soil particles behind the brush mat fill any small voids thus providing favourable conditions for rapid root development. The rapid formation of a dense wall of branches will protect and stabilise the bank reliably and permanently.

Advantages
Simple and quick method using local live materials. Repair work and maintenance do not pose any problems.

Disadvantages
Construction only during the dormancy period to early growing season.

Costs
Very much lower than for comparable solid structures.

Areas of use
Protection and erosion control of the vertical faces of slump failures of limited height where the use of solid structures would not fit into the landscape.

Branch layers

Oizinger developed a method for the repair and restoration of very steep or even undercut bank failures on fine grained sandy–silty soils, which resembles the brush structure by Prückner. A row of piles is driven into the stream bed along the line of the original bank alignment. The space between the piles and the face of the failure is tightly packed with branches of shooting and non-shooting shrubs and trees, providing immediate protection to the damaged bank. The subsequent rooting and shooting of the live branches will further aid in stabilising the endangered bank section, leading ultimately to the establishment of a plant succession adapted to the local conditions.

Timber cribs and crib walls (Fig. 3.31; Plates 47–49)

Cribs and crib walls as an element of transverse structures were described in Sections 3.3.1.3 'Wood sills' and 3.3.1.4 'Crib walls'. They can also be used to great effect in longitudinal training measures as revetments.

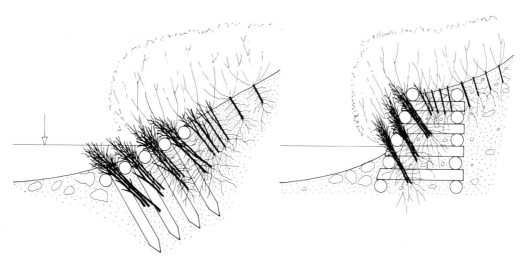

Fig. 3.31 Single and double timber cribwork training walls reinforced with live cuttings and branches.

Materials
Timber poles or cut timber; alternatively, if necessary, more expensive prefabricated sections in steel or concrete; nails; bolts; gravel; crushed stone; quarry stone; live branches of trees and shrubs capable of vegetative propagation (see Tables 2.5 and 3.5, at ends of Chapters 2 and 3, respectively).

Implementation
Round timber is preferred, because its use permits the construction of curved crib walls that fit the terrain. A prominently raked wall will fit better into the natural landscape than a steep or vertical structure.

The live branches must reach through the crib and into the soil of the bank to ensure rooting. They must not be packed too closely and bedded in soil for their total length in the crib in order to facilitate rooting over their whole length. It is of advantage to place the branches at an angle of 25° sloping backwards. The tips of the branches should protrude some 0.25–0.5 m from the crib wallface.

Timing
During the dormancy period.

Effectiveness
Immediately very effective. Reinforcing action when taking root, resulting in the formation of stable shrub or small tree rows with mixed vegetation cover. Ultimately, the established vegetation will completely take over the function of the crib wall when the timber frame, after many years, finally decays.

Advantages
Simple and quick construction. Materials may be obtained from nearby woodland.

Disadvantages
Construction strictly only during the dormancy period. Poor results are obtained when live branches and cuttings are planted after the construction of the wall. Treated timber must be transported from further afield.

Costs
Low, if timber and branches available from local source.

Areas of use
Bank protection of torrents and mountain streams with seasonally fluctuating water levels and base load as, for example, in limestone areas of the Alps.

Live timber frames (Plate 50)

These frequently adopted timber-frame structures for protection and stabilisation of the earthworks are equally effective for the repair of damaged stream banks and banks of dams, etc. For the latter purpose, they are often used as substitutes for high-cost and labour-intensive criss-cross wattle fences.

Materials
Treated or untreated timber poles; alternatively, prefabricated and more

expensive sections of steel, concrete or synthetic material; bolts and screws; nails; straps; free draining backfill and topsoil; live branches and/or cuttings; rooted plants; grass and legume seed.

Implementation
The timber framework is placed on the slope and, after reinforcement with live branches, filled with free draining soil. Double (three-dimensional) frames require large amounts of live materials to reinforce the whole structure with live brush layers, cuttings, etc.; in addition, a quick growing ground cover of grass is usually established. It is important to provide a strong base for the frame to avoid buckling or collapse.

Timing
If live materials such as branches or cuttings are used, only during the dormancy period. Seeding of grass or turfing, planting of container or bare-rooted plants during the growing period.

Effectiveness
Live timber frames prevent erosion and stabilise the bank immediately after construction is completed. The roots of the growing vegetation cover increase the effectiveness of the structure and take over its function.

Advantages
Many variations and combinations with other methods are possible. Immediate effectiveness. Landscape appearance not disturbed.

Areas of use
Banks of fast flowing rivers and lake shores subject to wave attack. For the erosion control and rehabilitation of steep slopes which, due to lack of space, cannot be modified. Repair of bank failures. The maximum height of timber frames is limited to approximately 15–20 m.

Cylindrical wire baskets

Also known under the name of gabions, their construction is described in Sections 3.3.1.3 'Live wire mesh gabion sills' and 3.3.1.4 'Wire gabion barriers', and the same criteria apply to their use in longitudinal bank protective measures. It is essential that the live cuttings or branches are firmly embedded in the soil of the bank to ensure adequate rooting.

The gradual disintegration of the wire mesh is counteracted by the stabilising effect of the growing plant cover. This effect can be enhanced by the choice of suitable plant material at the time of construction to allow for the formation of secondary and permanent plant communities. Cylindrical wire baskets are mainly used in the southern Dolomites of the Alps and the karst areas.

Geotextile or geogrid reinforced structures (Plates 51 and 52)

Geotextile or geogrid sheets manufactured from durable synthetic fibres wrapped around sand, gravel or soil are used for the construction of a variety of different shaped elements, ranging from rectangular or square blocks to long cylindrical or flat mattress-like structures. The variability and flexibility of such structures makes them ideally suitable for the construction of large and stable training and support structures (Zeh, 1982). Geotextiles and geogrids (also collectively termed geosynthetics) are the modern equivalent of sandbags which were effectively employed for the control of catastrophic flash floods.

Similar to the reinforcement of cribs and gabions with live branches, geotextile/geogrid reinforced structures or layered blocks are usually constructed in combination with live vegetative materials. The same principles as outlined in Sections 3.3.1.3 'Live wire mesh gabion sills' and 'Wood sills' and 3.3.1.4 'Crib walls' and 'Wire gabion barriers' apply. The advantage of geotextile (and geogrid) reinforced structures is that live cuttings, branches, stakes, etc. may not only be placed between the individual structural elements, but may also be inserted into them by pushing them through slits cut into the geosynthetic layers. This has the advantage that the basic stability of the structures are not weakened by this process. Moreover, the development of the vegetative cover hides the unsightly shapes of geotextile and geogrid reinforced fills.

Geotextile reinforced soil structures with vegetative cover are flexible and ideally suited for the protection of stream banks and river training; they are resistant to fluctuating water levels and heavy bed loads. However, their relatively high cost as compared with other structures with similar efficiency is a disadvantage.

Flexible bank protection method (Watschinger and Dragogna, 1968) (Plates 53 and 54)

This method has been used with great success for the last 25 years in South Tyrol for torrent control. It does not disturb the landscape and can, on account of its flexibility, cope with fluctuating water levels and bed loads.

The toe of the bank is secured by a line of large machine-laid stones of 300–500 kg weight. The stones are held together by wire ropes, which in turn are secured to a series of stout timber poles driven into the bed. If large stones are not obtainable, wire gabions are used instead. To prevent the stone rows from sinking into the ground, they are placed on stone sills which in turn may rest on a line of piles (in soft stream beds). Live willow branches or cuttings are placed between the stones or in the gabions, and the adjacent sections of the bank are protected by brush mats or closely spaced live cuttings.

This system of flexible bank protection is based on a formerly much-used method known as the 'Rosary' among Austrian river engineers.

3.4 Supplementary construction techniques

Supplementary construction techniques are undertaken with the purpose of diversifying the originally established vegetative cover and safeguarding its continued existence in order to achieve the objective.

3.4.1 Planting of pot plants and containerised plants

The transplantation of bare rooted plants to revegetate earthworks, etc. has rarely been successful in Alpine areas because the prevailing site conditions are usually too extreme. This method is therefore best reserved for the afforestation of areas with a suitable climate and adequate soil.

Seedlings and plants with their roots firmly covered by the original planting medium have a better chance of survival on transplantation than open rooted plants. Raising herbs and grasses (but not woody species) in containers for transplantation in areas where seeding is of doubtful outcome has proved to be equally advantageous.

Implementation
Sleeved or container plants are planted in prepared planting holes. Depending upon local conditions, the planting holes are made by hand or mechanical means. Tractor-drawn planters are only suitable for use in relatively flat areas on level ground; on steep slopes planting holes have to be dug by hand or a suitable soil auger.

In many instances it will be necessary to remove larger stones from the planting holes and replace them with topsoil or compost. To encourage rapid growth, it is of advantage to keep the immediate vicinity of the planting station free of weeds and other competing vegetation. This is best achieved by covering the station with an organic mulch or plastic foil. The practice of chemical weed control is to be avoided. Watering the stations after planting will ensure rapid growth.

Timing
Throughout the year, except during periods of frost. The most suitable time is the transition period from dormancy to the growing season.

Costs
Very variable depending upon site conditions and plant material; no indications can therefore be made. 700–1200 planting stations per two man team per day. 10–25 plants per hour if the planting holes have to be dug by hand; 25–60 plants per hour if a planting auger is used.

3.4.2 Transplantation (Plates 2, 3 and 55)

In this context, transplantation techniques involve the transfer of large and well-bound dense topsoil sections complete with their vegetation cover from nearby locations to the construction site (Horstmann and Schiechtl, 1979).

Materials
Rooted topsoil with its plant community of turf, herbs, dwarf shrubs, large shrubs or even small trees; care should be taken to disturb the root mass as little as possible.

Implementation
The sections are lifted by back hoes of similar mechanical equipment and transported to the site by front-end loaders, where they are placed immediately into prepared shallow excavations.

Timing
For best results, during the dormancy period.

Effectiveness
The transplantation of large topsoil sections with their plant communities creates ecological islands or focal points in otherwise bare areas, facilitating the re-establishment of the local fauna and flora within a relatively short time.

Costs
Only economical if the vegetated transplant sections can be obtained from suitable nearby sites.

Areas of use
Selective rehabilitation of large areas of bare ground if large topsoil
sections can be obtained from sites within the construction zone and the
necessary mechanical equipment is locally available.

3.4.3 Planting of root stock divisions

Materials
Multi-stem plants and tufts of grasses, herbs and, in some instances, also
woody species that can be divided, such as brome grasses (*Brachypodium*
spp.) yarrow (*Achillea millefolium*), tufted hair grass (*Deschampsia
caespitosa*), sedges (*Carex* spp.), etc.

Implementation
Suitable plants are dug up, either in their natural habitat or in special
nurseries, and divided into sections large enough to ensure ready
rerooting. It is of advantage to add some fertile topsoil and/or compost
to the planting station. When transplanting short grasses in rather arid
localities, the addition of topsoil is not necessary.

3.4.4 Rhizome plantings (Fig. 3.25)

Materials
Live rhizomes of suitable plants, e.g. reed (*Phragmites*), iris (*Iridaceae*),
butterbur (*Petasites*), reed mace (*Typha*), sweet flag (*Acorus calamus*),
etc. The length of the rhizomes and planting density depend upon the
selected species and the desired effect; as a rule, three to five pieces per
sq m are sufficient.

Implementation
Single rhizome pieces or root crowns 100–150 mm in length are placed
into shallow planting holes and lightly covered with soil. Long and
straight pieces of rhizomes can be planted vertically like live cuttings,
leaving only a short length protruding above ground level. On poor and
stony–gravelly soil of poor fertility, some fertile topsoil or compost
should be mixed with the planting soil.

Table 3.3 Soil protection techniques.

Technique	Areas of use	Suitability and effectiveness	Advantages	Disadvantages	Timing	Costs
Sods and turf	Protection of erodible areas	1 All areas where natural sods and turf are available	+ adapted to local conditions + immediately effective + simple, quick method	– difficult to obtain	Growing period	Mechanised: low Hand labour: average
Rolled turf*	Streambanks, waterways, landscaping	2	+ immediately effective	– only on topsoil	Growing period	Low to moderate
Grass seeding: (a) Hayseed seeding	High altitude sites, nature reserves, combined with other methods	2–3 In combination with other methods	+ adapted to local conditions + Multi-species community	– difficult to obtain – only on topsoil	Growing period	Low
(b) Standard seeding*	On topsoil, permanent or temporary cover	1 On topsoil with no erosion hazard	+ quick, low cost method	– only on fertile topsoil	Growing period	Low
(c) Hydroseeding**	Mechanised seeding of steep slopes, on subsoil	2 In shady locations in humid climate	+ quick, easy method + mechanised + spreading of all components in one operation	– vehicular access required – limited reach – moderately successful on dry sunny slopes	Growing period	Low to moderate

Technique	Application	Advantages	Disadvantages	Suitability	Best period	Rating
(d) Mulch seeding**	Protection of large areas in cuts or on fills on subsoil	+ creates micro-climate + quick successful germination and growth (glasshouse effect) + formation of organic matter + mechanical protection of soil surface	– several work processes – slow decomposition of mulch at high altitude sites	1 All sites on subsoil	Growing period	Moderate
Direct shrub and tree seeding	Woodland establishment and supplementation of other vegetative methods	+ economical + natural + for areas where planting is impossible	– slow development	1 Rocky, very steep sites	Beginning or end of growing period	Low
Seeding on erosion control netting	Very steep slopes, sandy slopes, stream banks	+ immediately effective	– expensive	1–2 Erosion control	Growing period	High
Seed mats	Grassed waterways, gentle, even slopes	+ immediately effective	– needs fine seedbed, preferably on topsoil	1–2	Growing period	Moderate to high
Precast concrete cellular blocks	Parking areas, driveways, protection of watercourse banks	+ immediately effective + planting during use possible	– labour intensive – limited height of slope	2–3	Growing period	High
Live bush mats	Protection of slopes subject to run-off and wind erosion	+ immediately effective + fast growing dense bush cover	– high demand of live material – limited slope height	1–2	Dormancy period	Moderate

Suitability: 1, very good; 2, good; 3, limited.
*, on topsoil only; **, on subsoil or inert soils.

Table 3.4 Ground stabilisation techniques.

Technique	Areas of use	Suitability: ecologically	technically	Advantages	Disadvantages	Timing	Costs
Cuttings	Protection and erosion control of earth banks, stone pitching and dry stone walls	2–1	2	+ simple, quick method + can be implemented at a later stage	none	Dormancy period	Very low
Wattle fence	Stabilisation and retention of topsoil	2	3	+ immediately effective	– large quantities needed – limited effect in subsoil – limited rooting – sensitive to slumping	Dormancy period	Moderate to high
Layering (a) Hedge layer	Protection and erosion control of riverbanks and watercourse banks and land slides	1	2	+ immediate establishment of climax type vegetation + immediately effective	– needs fertile topsoil – large quantities of rooted plants required	Growing period	Low (moderately high*)

(b) Brush layer	Protection and erosion control of riverbanks and watercourse banks, and land slides	2	1	+ simple mechanised construction + subsoil stabilisation + use of spreading branches	none	Dormancy period	Low (low*)
(c) Hedge–brush layer	Protection and erosion control of riverbanks and watercourse banks, and land slides	1	1	+ establishment of pioneer and climax vegetation in one operation + effective at subsoil level	none	Dormancy period	Low (low*)

Suitability: 1, very good; 2, good; 3, limited.
*in comparison to individual stabilisation measures.

Table 3.5 Combined construction techniques.

Technique	Areas of use	Suitability: ecologically	technically	Advantages	Disadvantages	Timing	Costs
Branch packing	Gully control	2	2	+ long lasting effect	– large quantities of branches required	Dormancy period	Average
Palisades	Gully control	2	2	+ fast construction + immediately effective	– limited width and height – only on fine textured soils	Dormancy period	Average
Live brush sills or submerged weirs							
(a) Brush sills	Transverse structures in small streams	2	3	+ simple construction + flexible + easy to repair + combination with other methods	– not very stable	Dormancy period	Low
(b) Willow fascine sills on brush layer	Transverse structures in small streams	2	3	+ simple construction + immediately effective + combination with other methods	– not very stable	Dormancy period	Low
(c) Wattle fence sills	Transverse structures in small streams	3	3	+ immediately effective + combination with other methods	– large quantities of material required – limited effectiveness below surface – limited rooting – sensitive to bed load – labour intensive	Dormancy period	Average to high

Suitability: 1, very suitable; 2, suitable; 3, limited suitability.

				+ (advantages)	− (disadvantages)		
(d) Live wire mesh gabion sills	Torrents and mountain streams up to 10 m wide	3	2	+ simple construction + immediately effective + local materials + permeable	− labour intensive − difficult to establish vegetative cover after construction	Dormancy period	Average
(e) Live geotextile cylindrical sills	At lower reaches, up to 5 m bed width	2	2	+ simple construction + immediately effective + flexible + easy fit to terrain	− limited experience with regard to erosion resistance	Dormancy period	Average to high
(f) Wood sills	Gully control	2	1	+ locally available materials + simple construction + effective after short time + flexible	none	Dormancy period for live materials	Average
Live barriers							
(a) Crib walls	Bed stabilisation and sedimentation in torrent control	1	1–2	+ local materials + quickly effective + flexible + permeable	− placing of live materials after completion not possible	Dormancy period	Low to average
(b) Wire gabion barriers	Bed stabilisation and sedimentation in torrent control	2	2	+ local materials + immediately effective + permeable	− placing of live materials after completion very difficult or impossible − mainly in the Dolomites	Dormancy period	Average
(c) Dry stone wall barriers	Bed load retention	2	1	+ local material + simple construction + immediately effective + very stable	− construction with live material only during dormancy period	Dormancy period	Average

Technique	Areas of use	Suitability: ecologically	technically	Advantages	Disadvantages	Timing	Costs
Groynes, dykes and spurs	Deflection of fast flowing water; stream bank modification	1	1	+ ecological benefits + hydraulically effective structures	– requires large space	Mainly during the dormancy period	Average
Live brush sills	Bed siltation	1	3	+ simple and quick construction + combination with other structures	– not very stable – only in shallow water	Dormancy to early growing period	Very low
Brush traverses	Groyne deflector reinforced with live materials; rehabilitation of damaged stream banks	1	2	+ simple construction + fully effective soon after construction + very economical + may be expanded at later stage + simple maintenance + combination with other structures	– not suitable for rivers with heavy bedloads	Dormancy period	Low to average
Brush grids	New stream bank alignments; repair of extensive bank failures	1	2	+ simple construction + fully effective soon after construction + combination with other structures	– very labour intensive – large quantities of live materials required	Dormancy to early growing period	Average
Reed single stem or culm plantings	Slow flowing or still water; lake shores	1–2	2–3	+ simple construction + low cost	– limited construction period	Early May to mid June	Low

Technique	Application		Advantages		Disadvantages	Construction period	Cost
Reed clump plantings	Stream bank protection with moderately fast flows; lake shores	1	+ local materials + simple construction	2	– limited construction period – fully effective only after 2–3 years	Dormancy to end of April	Low to average
Reed rolls	Protection of stream banks, lake shores, and large drains	1	+ immediately effective + dense vegetation cover	1–2	– limited construction period – labour intensive – large quantities of material	Dormancy period to early spring	Average to high
Stone revetments reinforced with cuttings	Reinforcement of massive revetments	1	+ simple construction + creation of riparian woodland	1	– vegetative cover only fully effective after 2–3 years	Dormancy period	Average
Branch layers	Steam bank protection with moderately fast flows; lake shores	1	+ local materials + resistant to heavy bed loads and ice drift	1	– planting of cuttings after completion not possible	Dormancy to early growing period	Average to high
Live fascines	Protection of stream banks and lake shores	1–2	+ simple construction + immediately effective + combination with other methods	2	– construction only during the dormancy period	Dormancy period	Low to average
Branch packing	Repair of bank failures and bed erosion	1	+ stable structure + immediately effective + simple construction + local materials	1	– labour intensive – large quantities of live materials needed	Dormancy period	Average to high

Technique	Areas of use	Suitability: ecologically	technically	Advantages	Disadvantages	Timing	Costs
Live brush training structures	Protection of stream banks	1	2	+ simple and quick construction + easy repairs + combination with other methods	– limited construction period	Dormancy to early growing period	Average to high
Branch layers	Repair of bank failures on fine grained sediments	2	2	+ simple construction + local materials	– large quantities of live and dead plant materials needed	Dormancy period or whole year if used immediately	Average to low
Timber cribs and crib walls	Protection of mountain stream banks and torrents	1	1	+ simple construction + local materials in wooded areas + fits the terrain + flexible + immediately effective + permits free drainage of the bank	– limited construction period because later insertion of live material problematical	Dormancy period	Average
Live timber frames	Protection of steep and high river banks	1	2	+ immediately effective + area support + non visible + combination with other methods + subsequent vegetative cover	none	Woody plants during dormancy; grass cover and herbs during growing period	Average

Cylindrical wire baskets	Bank protection of torrents	2	+ local materials + immediately effective + flexible + permeable	1	− vegetation limited to pioneer types − placing of cuttings, etc. after completion of works not possible − limited construction period	Dormancy period	Average
Geotextile or geogrid reinforced structures	Protection of stream banks and shorelines	1	+ simple construction + flexible to fit the terrain + immediately effective + good vegetative cover of bank	1–2	− limited experience as to deterioration with time	Live woody plants during dormancy; grass cover and rooted plants during growing period	Average to high
Flexible bank protection method	Protection of banks on fast flowing mountain streams and torrents	2	+ very flexible + immediately effective	1	− labour intensive	All year round	Average

Timing
The propagation of plants by rhizomes is governed by a certain growth cycle which is specific to the various genera. This cycle is, however, not so pronounced as is the case with cuttings of woody plants. Rhizomes planted during the dormancy period give the best results; they should be planted immediately, as only short time storage in a cool place under a light cover of moist sand is possible.

Advantages
Simple method to establish a quick growing vegetative cover of plants for which seed is not available from commercial outlets.

Costs
Depend largely upon collection site, plant species, desired density and site conditions, but are rather high.

3.5 Special structures

3.5.1 Branch packing (Fig. 3.6, lower part)

Small and shallow gullies may be repaired by placing branches or bundles or branches, including those from conifers, butt end upstream, into the gully and placing stones on top to secure them.

Deep and steep gullies are filled completely with tightly packed branches, preferably from coniferous trees with the cut ends at the bottom (Fig. 3.6). The branches are kept in place by stout cross poles which are anchored into the sides of the gully; the branches nearest to the poles are tied with strong wire to the cross pieces. For economic reasons, this method is usually only used on rather narrow V-shaped gullies on account of the large quantities of branches required.

The branch packing effectively brakes the force of the fast running water, retaining its sediment load and permitting the water to drain freely away, leading in time to the complete siltation of the gully. Live branches and cuttings, placed with the pack, will lead to a permanent vegetative cover.

3.5.2 Anchored tree mats

Large multi-branched trees or tree tops are placed on bank failures as a first measure to stop all further erosion during high flood periods. The trees and tops are tied together with wire ropes or secured with large stones or piles to keep them in place.

Tree mats reduce flow velocity, thereby causing sedimentation of the bed load and retention of floating debris. However, they are only effective as long as the branches remain elastic and flexible. As soon as the immediate danger of further erosion is eliminated, more conventional engineering measures must replace the temporary tree mat to repair and secure the bank.

Submerged tree mats are placed into deep scour areas and weighed down with concrete blocks, rocks or old pipes, etc. to prevent them from floating to the surface. In most instances, submerged mats are soon covered with sediment.

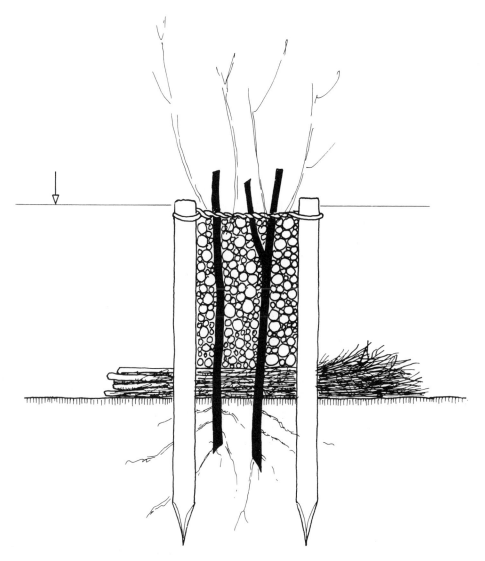

Fig. 3.32 Cross-sectional detail of a brush dyke.

3.5.3 Live brush dykes (Fig. 3.32; Plate 56)

Dykes are usually constructed in tidal areas to push the sea back and in the process gain agricultural land. More recently, some form of dykes are used to protect the banks and sides of large water-filled quarry or borrow areas against further erosion by excessive wave action (Plate 56). Between the shore and the dyke a belt of reeds is planted, which, protected by the dyke from wave damage, will soon establish itself and protect the vulnerable sides. Live dykes are made up of two rows of piles, and the resulting space between is tightly packed with branches of deciduous or coniferous trees. Opposite pairs of piles are wired together to stabilise the structure. The tightly packed mass of branches between the two rows of piles effectively dissipates the energy of the waves.

It is not possible to lay down firm dimensions for dykes. In one instance the piles used to construct a dyke were pine logs of 200 mm diameter and the coniferous branches were tied with galvanised wire. In another instance the stakes for a dyke to protect the sides of a large water-filled borrow pit were cut 30–50 mm thick.

The principal cross-section of any given dyke depends upon the local conditions. The limits are set by the dyke's ability to effectively diminish the energy of wave action. Height and width of the branch packing are of equal dimension. The actual size can only be determined on the basis of results of practical on-site or 'field' trials.

Brush dykes are only temporary measures which facilitate the establishment of other measures to take over the protection of any given shoreline or riverbank.

The difficult, labour intensive and very costly wattle dyke designed for the speedy siltation of narrow but deep scours, is very rarely if ever used nowadays in the Alps.

Chapter 4

Bioengineering Techniques in Earth Dam and Floodbank Construction

As permanent structures, earth dams and floodbanks must fulfil several functions. The materials used for their construction are subject to many forces, both external and internal, such as the translocation of particles and formation of new aggregates, oxidation processes and others.

The main function of any dam or floodbank is the storage and retention of water. Differences between the various types of dam structures are *inter alia* determined by the site conditions and the wetting periods of the upstream side of the structure during the annual cycles.

❑ Flood control dams or floodbanks – temporary wet conditions; pronounced fluctuation of the water level.
❑ Banks of navigable canals – permanently wet; generally constant water level.
❑ Reservoir dams – generally permanently wet; relatively small changes in the water level except during periods of increased run-off (high flood periods).
❑ Storage dams on permanent rivers feeding hydro-electric projects – wetting cycle as above.
❑ Storage dams on rivers with seasonally pronounced fluctuating run-off – constant wetting during periods of full supply; moderate variations during operation; minimal wetting during periods of maximum draw down.

As a consequence of the above points, there is only limited scope for the use of bioengineering techniques for the protection of the upstream side of river storage dam structures serving hydro-electric schemes, and no possibility at all for their use on the upstream side of the bank on compensation reservoirs.

If we accept the rather subjective and somewhat questionable dictum that the upstream faces of dam structures have to be kept free of all vegetation and that firm rules apply in this regard to the downstream faces of these structures as well, the following questions still arise:

❏ What is the effect of vegetation on the safety and projected life of any dam?
❏ Is dam safety and projected life of the structure increased or placed at risk by the vegetation?
❏ Is it possible to keep earth dam faces permanently free of spontaneous plant invasion?
❏ Should a structure be kept free of spontaneous plant growth anyway?
❏ Are bioengineering techniques designed to increase dam safety and visual appearance feasible, necessary or desirable?
❏ What would be the scope and the limits of bioengineering techniques in dam construction?
❏ What part of the dam structure and its immediate environs would benefit from bioengineering techniques?
❏ What and where are the limits for the application of bioengineering techniques?
❏ Which plants or plant communities are acceptable on the faces of earth dams?
❏ Do bushes and trees constitute a hazard to earth dams?

Bioengineering techniques have been used for some time for the protection of flood control dams, reservoir dams and canal banks. What limits or prevents such techniques from being used on other storage works? Possible reasons are:

❏ The use of asphalt–cement covers.
❏ The protection of the upstream slope by exclusively hard measures.
❏ The anticipation that the root system of established vegetation may cause serious deterioration of the filter layers, which in turn may jeopardise the whole structure.

There are, however, instances where these restrictions or prohibitions were ignored by enterprising civil engineers and vegetation covers were established, particularly on the upstream slopes.

❏ Banks of canals in France were planted in some instances with tall *Populus pyramidalis* before the turn of the century without deleterious results.
❏ Flood protection dykes on the river Elbe in the reach between Hamburg and the border of Lower Saxony were constructed to various designs. In Hamburg, the slopes facing the canal were finished with a layer of asphalt; in Lower Saxony they were vegetated, which has since developed into woodland. Both versions have withstood repeated very high flood levels.

❏ The downstream slopes of the earth structures of the river storage dams on the river Drau in Carinthia, Austria, have been planted to trees and bushes since the 1960s without objection by the responsible authorities (Kelenc, 1986).

❏ Soon after 1986, dam face zones between full supply level and maximum draw down were covered with rip-rap in combination with vegetative cover; this was subsequently extended to include the upper sections. So far, the results are very encouraging.

❏ All earthworks and banks of the Oberaudorf–Ebbs hydro-electric project on the river Inn to the north of Kufstein were extensively protected by the use of bioengineering techniques between 1990 and 1993.

❏ Following a directive from the Nature Protection Authority, between 1990 and 1992 the downstream side of the earth dam serving the Koralm–Feistritzbach–Soboth hydro-electric project of the Electricity Supply Authority in Carinthia was covered in rip-rap in combination with bioengineering techniques. These included erosion control measures and ground cover works involving turf transplantations and planting of woody species.

The upstream face of earth dams are not only subject to the erosive forces of natural precipitation and temperate changes, but also to attack by wave action. Under normal circumstances, the flow velocities within the storage zone of river hydro-electric schemes are rather low, but during high flood periods or basin flushings, high velocities are created and fluctuating water levels of a moderate order are encountered. Wave action is usually light to moderate, but persistent winds may subject the water line to severe attack. The banks of navigable rivers are subject to severe wave attack, the receding water causing serious sapping erosion resulting in loss of fine soil particles.

In the long run, such forces are better met by the use of vegetative or 'soft' means as opposed to purely mechanical methods or 'hard' structures. A dense, multi-layered mass of flexible branches, with or without leaves, will dissipate the energy of moving water very effectively. The root system will consolidate the ground and the mass of fibrous fine roots will bind the fine particles. Suitable plants will reduce excessive soil moisture through the process of evapotranspiration.

Plant roots have a physiological and mechanical function, such as the uptake of nutrients, water and oxygen, and the consolidation of the soil. To achieve this objective, a diverse system of sensitive fine fibrous roots and a strong network of large supportive roots is required. In order to tap the resources of nutrients and moisture, the roots of different plants will penetrate to variable depths. The depth of root penetration depends upon the degree of soil compaction: the tips of the growing roots, even those of woody species, are extremely sensitive and grow along the line of least resistance, avoiding any obstacle in their path.

In practical terms, grasses and herbs require an effective soil depth of 0.25 m, shrubs a minimum of 0.5 m and trees at least 1 m. Under natural conditions even deeply rooted trees rarely penetrate to greater depths than 1–2 m. Roots that penetrate to greater depths invariably follow zones of weakness such as cracks and fissures in the parent material or substratum that are filled with fine soil.

It follows that the upstream face of earth dams may be planted to vegetation provided that accepted construction parameters are adhered to. Up to the present time, not a single case of damage to the core, when correctly constructed with the proper materials, caused by root penetration, has come to light. The core constitutes an effective root barrier.

All earth dam surfaces must be protected against erosion and infiltration by surface water. An efficient ground cover vegetation (turf) will fulfil this function.

The downstream side of the dam will always be subject to invasion by woody plants; an occurrence which should not be restricted in any way. In principle, there is no valid argument against the establishment of a vegetative cover on the downstream side of the earth dam; moreover, its very extent invites such treatment.

From the structure's safety point of view, the proper drainage of those dam face zones that protect and support the core zone is of great importance. Under no circumstances must plant roots interfere with this drainage function. They may, however, augment the drainage process through normal plant water use during the growing period, relegating the drainage function in the dam face zones to lower strata (Hartge, 1986).

There is no room for tall coniferous trees on dam structures. Even moderate winds cause such trees to oscillate, transmitting the resulting forces through the root system to the soil which may cause unacceptable loosening of the surface structure. Strong winds increase the whipping motion of the trunks causing windthrow and fractures. Such windthrows are dangerous scars which lead to serious erosion and need immediate remedial measures to prevent further damage. Natural conifer volunteers on dam structures therefore should be removed as a matter of course or kept down by early felling to prevent the formation of mature trees.

The topping of dams and floodbanks by extreme flood waters is always a disaster because only in exceptional cases are these structures designed to withstand such events. There are numerous examples of the total or partial destruction of dam and floodbanks throughout the world.

In the event of flooding, single trees on the land side of a floodbank or dam may cause scouring. This can be avoided by managing volunteer shrubs and trees to form discrete clumps or groups and by arranging plantings in a similar manner, surrounding the trees with a small belt of bushes and shrubs (Fig. 4.1). It is not necessary to remove all trees from the site of a projected dam or floodbank. Those at the base may remain to be partially covered by fill if the maximum reach of the roots is 2 m short of the impervious core. To facilitate the survival of such trees, it is of advantage to complete the wall in several stages spread over two to

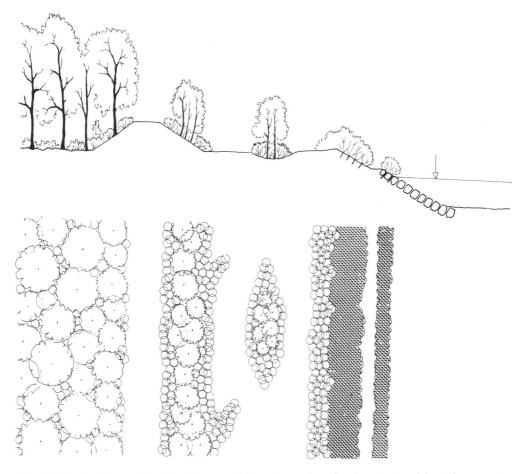

Fig. 4.1 Typical plan and profile of a revegetation scheme for a flood protection dyke with protective shrub planting around tree trunks.

three years. The trunks should be surrounded with a layer of pervious material (gravel, crushed stone), but never with clayey fill. This technique has saved over 300 specimens of white willow (*Salix alba*), grey alder (*Alnus incana*) and bird cherry (*Prunus padus*) at the site of the Oberaudorf–Ebbs hydro-electric project on the river Inn although they were covered up to a height of 6 m! Soon after completion of the earthworks, the formation of masses of adventitious roots could be observed (Plates 1, 57 and 58). It can be safely assumed that these trees contribute towards the stability of the land side of the floodbanks.

It is inevitable that shrubs and trees will get flooded in the basins of dams and weirs; their chance of survival is as a rule rather slim. However, it has been known for the past 50 years that various deciduous trees which share some characteristics of pioneer plants can withstand partial soil cover, temporary flooding or even permanent waterlogging conditions (Seifert, 1965, 1970; Schiechtl, 1973). Mainly willow (*Salix*), alder (*Alnus*), poplar (*Populus*), elm (*Ulmus*), ash (*Fraxinus*) and bird cherry (*Prunus*) belong to this group. Temporary flooding is tolerated even during the growing period. It has been observed over a number of years that willows and poplars withstood total submersion by the icy waters of the river Inn for periods of up to six weeks without ill effects, except some damage caused by floating debris.

Numerous observations and investigations have been used in the compilation of the following list of measures to be taken for the preservation of permanently waterlogged trees:

❏ Trees do not tolerate sudden permanent flooding. During the construction period, the base of the trunks should have been covered over several stages with a layer of pervious material. This will lead to the formation of a vigorous mass of adventitious roots over the covered portion of the trunk which will ensure the survival of the tree when the old root system has succumbed to permanent waterlogging.

❏ The longer the time period spent on the above mentioned preparations, the better the chances of success. The total length of time involved before flooding should be two to three years.

❏ The depth of gravel cover should always exceed the water level so that the trunk is never standing unprotected in water. For best results small groups of trees should be covered as a unit, resulting in green islands dotting the shallows (Plates 59–61).

❏ It is important that no fine textured materials such as silt, loam or clay are used for the cover of the trunks; this would restrict the oxygen supply to the newly formed roots, causing the death of the tree.

❑ If the soil mantle extends above the full supply level of the dam, it is best left for the natural establishment of flora and fauna. Plant succession of such habitats surrounded by water is more rapid than on dry land.

Ideas, experience gained and the basic concepts of the progressive relationship between water, engineering and landscape as applicable to river storage works and flood protection bunds, together with the need to work with nature, the use of bioengineering techniques and the care and maintenance of the vegetation established for the protection and erosion control of civil engineering works, have been published in many papers. It is therefore not the intention of the authors to reiterate the subject (Binder, 1986; Gilnrainer, 1986; Vitek, 1986). These publications specifically deal with the general layout and detailed design of dams, weirs and the renaturalisation of water courses together with aspects of their landscape design.

Chapter 5

Wetland Areas

The interrelationships between flowing and still water and wetland areas are manifold and often of a rather complicated nature. The following short notes should highlight the essential points.

The term 'wetland' comprises all those ecological systems that are characterised by their dependence upon a permanently high water table and/or surface water, such as back waters, reed zones, marshes, swamps and bogs, including alluvial woodland and bush, regardless of the altitude where they occur.

The economic use of such land was relatively limited in the past due to the extremely wet conditions and the difficulty of vehicular access. Possible exceptions were meadows and parts of alluvial woodland or forest. Owing to increased population pressure, large wetland areas were cut off from their natural water supply by artificial drainage, peat cutting, levelling and silting up of depressions and the construction of floodbanks. This interference often resulted in the destruction of the wetlands, and the few relics still found in Europe are remnants of formerly much larger areas. Wetlands are often the last refuges of many rare plants and the habitat of rare animals.

Wetland can only be maintained if the ecological base can be maintained – in the first instance, the water regime and nutrient cycle. Connecting channels to nearby water bodies must remain open and the free movement of ground water must not be restricted. On the other hand, eutrophication from neighbouring agriculturally used areas must be avoided and, if necessary, prevented by artificial means. Wetlands are very sensitive to toxic waste pollution originating from waste tips, mine dumps, industrial waste, dust, or even pollution caused by rain (acid rain).

Any environmental assessment for planning purposes should be based on a thorough botanical survey in order to pinpoint any future change in species composition and abundance, and examination of the nutrient content of the soil and its water regime, ground water levels, phreatic horizons and water quality. If the outcome of these investigations confirms that the wetland is worthy of preservation, the necessary steps must be taken to declare the area a nature reserve or similarly protected land. If it is unavoidable that the wetland has to give way to development, transplantation of plant communities can contribute to maintenance of the plant species. The complete destruction of any wetland either by soil removal or dumping should be avoided at all costs.

For the successful transplantation and re-establishment of wetland vegetation, the following criteria must be met:

❑ selection of a suitable area (Plate 62);
❑ creation of the required ecological conditions, e.g. soil, water supply and river flow, control of water levels, and nutrient content;
❑ protection from access by people, grazing animals, and eutrophication;
❑ natural turf or soil sections complete with their vegetation cover should be as large as possible (Plate 2).

If the transplants cannot be effected immediately, short-term storage should be planned and prepared very carefully. Total storage time must not exceed one growing period and the ecological parameters must be the same as those pertaining to the original site. Careful maintenance of the storage site is therefore essential.

Chapter 6

Care and Maintenance of Vegetation Along Watercourses

As stated earlier, it is a characteristic feature of bioengineering techniques that their effectiveness will increase with the passage of time as the plants grow and reach their full potential. To encourage growth and to shorten the time required to reach maturity, certain cultural practices and maintenance works are necessary. The more extreme and difficult are the growing conditions at the project site, the more intensive these must be.

Care and maintenance works are divided into three phases:

❑ Establishment phase
❑ Development phase
❑ Maturity.

6.1 The establishment phase

During the establishment phase, intensive cultivation and care aim at bringing the bioengineering techniques to the stage that is acceptable for commissioning. This stage is defined as follows:

❑ Seeding and seed mats: the specified grasses and herbs must form a uniform effective ground cover of at least 50%. Volunteer grasses and herbs of equal value are acceptable and are included in the assessment of the percentage ground cover.
❑ Seeding of woody species: seed distribution and mulch must be even. Seedlings of woody species and other plants contained in the specified mix must germinate evenly and in the proportions laid down.
❑ Natural turf and rolled turf: these must be well rooted so that they cannot be lifted off the ground.
❑ Planted trees and shrubs: failures must not exceed 30% of the total numbers planted, and the objective must be achieved in spite of the failures.

❑ Live plant materials: fascines, brush layers, hedge layers, hedge–brush layers and wattle fences must, on average, show five, and at least two, vigorous shoots per linear m. Brush mats must show on average ten, and at least five, shoots per sq m, evenly distributed. Two-thirds of all live cuttings and truncheons or stakes must have evenly distributed shoots.

These conditions may of course be varied in the tender specifications to suit the local conditions. Care and cultivation during the establishment phase also includes the replacement of dead plants and plant materials and other measures to promote vigorous growth, such as fertilising, irrigation, mulching, clean cultivation and weeding, together with plant protection, e.g. pruning, staking and tying, wound dressing, and disease control, and prevention of damage by animals and game.

6.2 The development phase

During the development phase, all cultivation measures are aimed at achieving the contractual objective. On average this takes between two and five growing seasons and ends with the commissioning of the project which is by then fully functional.

Operations may need to be undertaken during the development phase as set out in the following sections.

6.2.1 Fertilisation

As the objective of the fertiliser applications is aimed at securing optimum vegetative growth and not yield, the application rates must be calculated accordingly. The revegetation of earth works in civil engineering is usually confined to inert and infertile subsoils or similar material, and the supply of plant nutrients is therefore of great importance. Experience has shown time and again that the application of fertilisers accelerates the spontaneous establishment of plants on such soils to a marked degree. Fertilising accelerates the formation of an effective ground cover, thereby reducing the time span during which the structures are at risk. On poor soil, regular applications of fertiliser are therefore essential for the rapid development of the plant cover. The type of fertiliser used and the quantities applied depend largely upon the site conditions.

Four main methods of fertilisation and their combination, if required, should be considered:

- ❑ The application of mineral fertilisers.
- ❑ Manuring or other organic fertilisers.
- ❑ Composting.
- ❑ Green crops.

6.2.2 Irrigation

In temperate climatic regions, irrigation should be applied sparingly and limited to achieve certain objectives (soil consolidation, easy germination and plant establishment, countering drought periods). In areas with a generally humid climate but with a pronounced summer dry period, some supplementary irrigation may be necessary. In semi-arid areas, the establishment of a vegetative ground cover or other plantings without irrigation is not feasible (Lange and Lecher, 1989).

6.2.3 Soil cultivation

This is essentially required to loosen the soil and destroy weeds.

6.2.4 Mulching

Individual planting stations or the whole seeded area are covered with a layer (100–200 mm thick) of degradable material, preferably straw; mown grass or cut weeds may serve the same purpose. The mulch cover influences and, to a certain degree, modifies soil temperature and the moisture regime. This encourages plant growth and improves soil structure (Karl, 1990). To prevent the build up of excessive rodent populations, it is advisable to remove the mulch before the onset of winter.

6.2.5 Mowing

Mowing is not essential, although a single cut is generally recommended for all established grass covers. A cut stimulates stem and root growth and invigorates the growth of grasses and certain herbs, particularly if high seeding rates were used.

6.2.6 Pruning

During the first two years after planting, woody plants must be trimmed to shape and dead wood must be removed. Single shoot shrubs and certain trees can thus be forced to assume a bushy habit.

6.2.7 Staking and tying

Saplings of tall trees must be tied to stakes. Care must be taken not to damage the root system of the plant when driving the stake into the soil. Depending upon root and stem development, staking may be necessary for the first three to five years to support the tree. During this period, periodic checks must be made to ascertain and condition of the stakes and ties.

6.2.8 Plant protection and prevention of wildlife damage

The control of diseases and insects should be carried out by biological means wherever possible without the use of chemicals. Wildlife damage is usually due to browsing, grazing or rubbings to remove the velvet from deer antlers. The best protection, although expensive, is a fence; otherwise, game damage can be prevented to a certain degree by the use of chemical or other mechanical means, such as tree tubes or mesh guards.

6.3 Maturity

After the established vegetation has become effective ('maturity'), any further maintenance work depends on the site conditions.

Short-term maintenance comprises all those measures that are aimed at the protection of the established vegetation to ensure its continuing efficacy. These works are carried out either by a specially trained team of workers employed by the client, or by a specialist firm under contract. If the correct civil structures together with the most suitable bioengineering technique are implemented, the intensive maintenance phase will come to an end after two to three years; the vegetation will naturally develop into plant communities adapted to the prevailing conditions. This natural and self-regulating process will, however, take some time, which can be shortened through the implementation of special measures; shrubs and trees particularly require some measure of medium- and long-term maintenance.

The type of maintenance work to be carried out depends upon the plant cover and associations present, and their function:

❑ Permanent woodland to be used for the production of live material capable of vegetative propagation: periodic severe pruning and cutting back to the main stem; provision of cuttings, long stems, stakes, poles, etc.

❏ Permanent plant associations for the protection and erosion control of earthworks, etc., such as riverbanks, wind and noise protection: periodic pruning to maintain vigour and obtain live material; maintenance of efficiency.

❏ Climax vegetation, alluvial shrub and woodland: all maintenance operations will follow accepted silvicultural practices. The woodland will be mainly of the deciduous type of mixed woodland with conifers playing only a minor role. Felling and pruning should aim to establish optimum efficiency. On steep slopes inclined to slides, the establishment of mature and large trees is to be avoided by shortening the felling cycle. All silvicultural work will be done in cycles and it is of advantage to have this work carried out by trained forestry personnel.

Maintenance work is more intensive and carried out in shorter cycles if the established vegetative cover is, for engineering reasons, different to that which is natural to the local conditions, for example, if, for reasons of land use, economics or landscaping, grass or shrub is to be maintained in an area where forest is the natural climax.

It is of advantage to prepare a maintenance plan, as usually several bodies with diverging interests, such as the contractor and the agency responsible for project operation and maintenance, are involved. In the same manner as the construction of bioengineering techniques are tied to a certain time of the year, so are the various maintenance works. For large engineering projects, it is best to prepare a maintenance work plan for the whole year to make sure that no part of the task is left out or forgotten (Tables 6.1 and 6.2).

Water engineering measures whose main function is the stabilisation and protection of structures and areas require constant maintenance.

Table 6.1 Implementation schedule for maintenance work, valid for the temperate zone of the Northern hemisphere.

Month	Type of work
March–June	Repair work on turf and grassland.
May–September	Irrigation, tying, soil cultivation, fertilising, mowing, mechanical weed control, mulching, staking, fencing.
September–December	Prevention of wildlife damage.
October–April	Filling in gaps.
December–March	Cutting and pruning of woody plants: elimination of volunteers, thinning, rejuvenation.
All year round	Removal of unwanted species.

Table 6.2 Checklist of aftercare and maintenance works.

Establishment phase

Replacement of dead plants
- reseeding
- replanting
- replacement of plant materials (ground cover, turf, etc.) that have not taken

Promotion of growth
- fertilising
- irrigation
- mulching
- elimination of unwanted plants, weeding, mowing

Plant protection
- pruning of diseased branches and dead wood, wound dressing
- staking and tying of trees more than 1 m high
- fencing
- protection against wildlife damage by chemical means, shelters and guards, disease control, mainly fungi, insect control

Short-term maintenance

- fertilising
- mowing or controlled use as pasture
- soil cultivation, aeration
- rejuvenating pruning of woody plants
- thinning woodland and removal of unwanted species
- fence maintenance
- prevention of wildlife damage

Medium- and long-term maintenance

Silvicultural works
- thinning, radical pruning
- removal of diseased and dead wood
- pruning to obtain live material

Plantings at some distance from the riverbank on level ground, whose function is of secondary importance and which fulfil only a supplementary role, are largely left alone after the initial maintenance work to develop along the lines of the natural succession, to mature into riparian and alluvial woodland (Plate 63). Established natural woodland (Pockberger, 1952) requires its first management and maintenance work 20–30 years after planting. After a period of 50 years, woodland management and cropping will be of a more intensive nature to ensure the continuing multi-functional performance of the plantation. This will maintain the alluvial woodland and its protective function and enhance the appeal of the landscape. The diversity of such woodland in both the technical and biological sense is illustrated by its multi-purpose function. It entails, however, the obligation to carefully consider the double effect any interference will have on the woodland itself, and on the side effects with regard to run-off and soil stability, particularly during periods of above average rainfall.

The practical aspects of silvicultural measures and woodland management are nevertheless simpler than the theoretical implication would indicate. The structure of woodland is indeed very flexible in many ways: planning errors do not have those far reaching consequences that are inherent in exclusively solid structures, where disastrous floods may lead to total devastation.

Glossary

Adventitious growth New shoots or roots formed not from buds but from permanent plant tissue

Armour layer Coarse surface layer of stones forming a revetment; protects fine sediments/soils against erosion by currents, waves, seepages; usually underlain by fatter fabric; also natural deposits of gravels over surface of sediments

Bar Accumulation of sediment in a river formed underwater in floods subsequently exposed at lower flows

Bank Permanent side to river, its crest or top marked by the first major break in slope

Batter Slope of a bank given in degrees from the horizontal or as ratio of vertical distance to horizontal distance (or vice versa US)

Berm A level or gently sloping step formed in face of cutting or embankment from 0.5 m to 2 m wide

Bioengineering Engineered construction method which applies biological knowledge during the project process and which uses biological materials like seeds, plants, plant parts, vegetation pieces together with inert materials to protect and stabilise slopes of:

earthworks – embankments, cuttings and natural slope repairs – *ground* bioengineering;

riverworks – banks of watercourses, shorelines of lakes, etc. – *water* bioengineering.

See water bioengineering

Braided channel A wide watercourse into a number of entwined courses or channels – anabranches; the channels usually carry a large amount of detritus

Brushlayer Live stems or branch cuttings of vegetatively propagating shrub or tree species laid lengthwise, sometimes criss-cross style, across 1(V):6(H) battered terraces cut into slope, or along upper surfaces of successive horizontal stretchers or facing units of timber grids or cribwork; acts as immediate slope drain, shoulder reinforce-

161

ment, over time develops as live root soil reinforcement and solar pump

Canalisation   Artificial modification of watercourses by widening, deepening, straightening, embanking or lining with stone, concrete, geotextiles or geomembranes

Climax community   Final stage of plant succession, which remains in its structure at that stage unless basic changes of the environment, through climatic or other influences from the outside (pasture, fire, clearing, etc.), take place; artificial climax communities that remain at a certain stage of development through occasional or continuous interferences by man (turf in potential forest areas)

Colonisation   Successful invasion of a new habitat by a species new to the area

Community   A group of plants and/or animals living in the same habitat and interacting with each other

Conveyance   Flow capacity of a watercourse dependent on cross-section characteristics including friction created by bankside vegetation

Coppice   Broadleaved wood which is cut over at regular intervals to produce a number of shoots from each stool; also known as copse; (v) to cut the shoots from a stool so that more will grow

Corridor   Part of floodplain or land along watercourse extending from crest of bank to approximately 50 m either side

Cribwork   Boxes formed from interlocking timber, concrete or steel components laid in successive courses at right angles at intervals, backfilled with stone or soil and provided with live willow brushlayers in apertures between successive courses along exposed face

Culvert   Covered channel or pipe passing under highway, land or structure

Cutting (engineering)   Normally linear excavation below level of open ground for canal, railway, highway, pipeline or other site formation

Cutting (vegetation)   Live stem, branch or root of plant which propagates vegetatively

Dormancy   A phase of plant inactivity during winter or other period of adverse climatic conditions

Dyke   Artificial embankment formed along canal or watercourse or around low-lying area

Ecoengineering   Another term for bioengineering but without the associated overtones of medical and genetic engineering of that word

Ecology   The study of how living things relate to each other and to their environment. Also used loosely to describe the interrelationship, e.g. 'the ecology of the site'

Ecotone   Border area between habitats or types of vegetation which may in itself constitute its own habitat or vegetation zone

Erosion   Removal of surface soils and rocks by action of water, wind, frost, ice and extreme sun/heat; internal erosion leads to change of the earth structure and piping; closed vegetation is the best safeguard against erosion; extreme erosion is caused by catastrophes, e.g. floods, fires, earthquakes, etc.

Establishment   Measures which enable a possible unendangered, fast and good rooting and growth of artificially started vegetation, as for instance inoculation, composting, watering, wind and frost protection, snow cover, prevention of damage by game, fencing

Eutrophication   Nutrient enrichments of a habitat by natural or artificial means; leads to dense and uncontrolled vegetation growth

Faggot   Bundle of long branches or fascine (UK term)

Fascine   Bundles of straight stems or trimmed branches of 15–50 mm diameter live osier willow or dead chestnut, hazel or willow laid parallel, butts at alternating ends, to form 1–2 m long cylinders tightly tied at each end and in the middle with wire, polypropylene twine or rope, installed either longitudinally along toes of bank protection systems or laid criss-cross fashion to form crib-work structures in banks

Fetch   Direct horizontal distance along wind direction over which wind generates waves

Flora   Total entity of all plants living in any given area

Gabion   A stone filled box or tube formed from galvanised wire mesh panels or rolls having great flexibility and strength

Geogrids   Extruded or punched or woven two-dimensional orthogonal arrays of synthetic polymer ribs or fibre filaments with integral or stitch-locked nodes, usually flexible and strong

Geosynthetics   Generic term encompassing all synthetic polymeric materials used in geotechnical engineering and bioengineering; includes geotextiles, geogrids, geocells, geomembranes and geocomposites

Geotextiles   Durable high tensile strength synthetic construction fabrics used for separation, filtration, drainage, reinforcement and erosion control of soils and crushed aggregates; biodegradable fabrics are made from natural fibres such as coir, jute, flax, remie, etc. and are used primarily for erosion control, also as soil reinforcement in conjunction with brush layering (live gabions) or as short-term subsurface filters or as separators holding back soil behind geogrids in steep slopes pending establishment of vegetation

Groyne   Structure built out at intervals from river bank to deflect watercourse currents and protect bank against erosion

Gully   A steep sided erosion feature formed by downslope water action;

unstable and recently extended drainage channel that transmits ephemeral flow, has steep sides, a steep head scarp and width greater than 0.3 m, depth greater than 0.6 m; enlarges by bed scour, by head migration upslope and by side collapse

Habitat   The normal abode of a plant or animal; the recognisable area or environment in which an organism normally lives

Hard engineering   Approach to design and construction utilising inert, rigid and/or massive structures such as brickwork, masonry, concrete and steel – opposite of soft engineering

Higher plants   All plants excluding algae, fungi, lichens and mosses

Hydrograph   A graph indicating water level, velocity or discharge in a watercourse plotted against time

Inoculation   Method of artificially infecting shrubs and trees or herbaceous legumes with, respectively, mycorrhizae or rhizobium bacteria that normally live in symbiosis with them

Joint planting   Insertion of live stem cuttings into interstices of rip rap, blockwork or other inert armour units into underlying ground

Legumes   Herbs, shrubs and trees of the pulse family, as nitrogen gatherer because of specific root clod bacteria; good ground improvers; many form specially strong and deep-reaching roots and are good ground stabilisers; an important part of seed mixes

Levee   A raised bank formed naturally or artificially along a river or riverine flood plain (US)

Live cycle costs   Whole cost of a project including costs of feasibility investigation, design, construction management (including after-care, monitoring and long-term maintenance) and of eventual rehabilitation or replacement

Live-pole planting   Ground bioengineering technique comprising the installation by driving, or insertion into pre-drilled holes, of long live stakes, rods or poles of 1–2.5 m length of poplar or willow (or any plant which propagates from cuttings) at close centres for slope stabilisation purposes as a form of live soil nailing

Meander   Broad looping bends generally occurring in series in the lower reaches of a watercourse

Pollard   A tree which is cut at 2–4 m above ground level, then allowed to grow again to produce a crop of branches; (v) to cut branches from such a tree so that they will regrow for use as poles or stakes or purely for tree management in certain areas

Reach   A length of channel or watercourse

Reinforced soil   Mass soil structures incorporating layers of metallic, synthetic or natural materials as tensile reinforcement to facilitate construction of steep slopes and earth retaining structures

Re-naturising   Restoration of canalised watercourses or artificial landscape to conditions closer to nature

Riffle   Section of watercourse with locally steep gradient and shallow
turbulent flow usually associated with pools immediately downstream
in a sequence in gravel-bed watercourses

Rill   Shallow downslope erosion feature normally less than 0.3 m wide
and 0.6 m deep

Rip rap   Randomly placed loose rock armour against erosion

Riverworks   Engineering works involving construction of or repairs to
banks of streams, rivers, canals and edges of ponds, shorelines of lakes
and sheltered portions of estuaries

Root characteristics

Extensive rooters   Plants with a far- and deep-reaching root system.
The reason for reaching deeply might be the necessity for a strong
anchorage or deep ground water level (xerophytes); the penetrated
ground is very often several cubic m deep; in any case, it is always
many times the size of the root volume (e.g. *Petasites*, *Salix*, *Epilo-
bium*, etc.); the food-supplying roots (haustoria) are situated either far
away from the main root or very deep in the ground at the far end of
the main roots; plants with taproots also belong to this group

Intensive rooters   Plants with mainly dense, short-reaching and very
bushy root systems; these plants need large amounts of humus and
may also be deep rooters if there is a good supply of nutrient substance
and water in the top layers

Shallow rooters   Shrubs and trees whose roots are located mainly
near the soil surface

Taprooters   Plants with usually a single, geotropic main root, with
very few side roots of any size

Rugosity   The unevenness or roughness of channels in contact with
flowing water

Run up   Vertical height above still water level that wave will reach on an
inclined structure or bank

Scour   Localised erosion of bed material – rock, sand or deposited silt,
gravel at toes of banks, returning walls or bases of bridge piers – by
fast flowing watercourses

Seepage   Water movement into or out of watercourse banks

Sheetpiling   Interlocking timber, concrete or steel (usually) driven vertically to form bank protection or retaining walls along watercourses,
or groynes or weirs across watercourses.

Sill   A narrow ledge of hard strata protruding from bank diverting
current artificial submerged weir

Soft engineering   An approach to the design and construction utilising flexible, modular structures usually vegetated above mean water level – synonym for bioengineering

Spiling   Panel of thin stem live willows or freshly soaked hazel and willow woven around vertical willow or hazel stakes – usually short lived

Spillway   A structure which passes floodwater through over or around a dam

SSSI   Site of Special Scientific interest (UK)

Stream load   Quantity of solids transported by a watercourse per unit time in suspension by traction by saltation

Terrace   Ledge or step formed naturally in banks of large watercourses or artificially in cuttings or embankments, usually wider than 2 m

Terracette   Narrow ledge or step formed in slope of 0.5 m or less in width

Toe   Base of bank or wall

Tractive force   Shear force over perimeter of wetted surface induced by and acting parallel and in same direction as flow in watercourse

Turbidity   The pollution of water by dissolved or suspended solids – silts, muds, chalk, metals or crushed coal, etc.

Water bioengineering   Branch of bioengineering concerned with water engineering exclusively

Watercourse   Natural or artificial channel which conveys water

Waterway   Navigable sections of rivers, channels or canals

Glossary compiled by Geostructures Consulting, Edenbridge.

References

Anselm, R. (1976) Analayse der Ausbauverfahren, Schäden und Unterhaltungskosten von Gewässern. Mitt. Inst. Wasserwirtsch., Hydrologie und landw. Wasserbau, TU Hannover, 36, 11–190. *(Analysis methods of new construction, damage repair and maintenance costs for waters courses.)*

Binder, W. (1986) Beispiele zur Stauraumgestaltung aus Bayern. In: 5 Sem. Landschaftswasserbau, TU Wien, 7, 307–342. *(Examples of dam sites in Bavaria.)*

Bittmann, E. (1953) Das Schilf und seine Verwendung im Wasserbau. Angewandte Pflanzensoziologie, 7, 5–44, Stolzenau/Weser. *(Reeds and their uses in hydraulic engineering.)*

DVWK (1984) Ökologische Aspekte bei Ausbau und Unterhaltung von Fließgewässern. Merkbl. zur Wasserwirtschaft, 204, 188 S., Deutsch. Verb. f. Wasserwirtsch., P. Parey, Hamburg – Berlin. *(Ecological aspects of new construction and maintenance of watercourses.)*

Ehrendorfer, F. (1973) Liste der Gefäßpflanzen Mitteleuropas. 318 S., G. Fischer, Stuttgart. *(Plant names.)* *(List of container plants of central Europe.)*

Fargue (1862) In: Bretschneider, H. (1982): Gewässerbau. Taschenbuch f. Gewässerausbau, Verl. P. Parey. *(Hydraulic construction.)*

Felkel, K. (1960) Gemessene Abflüsse in Gerinnen mit Weidenbewuchs. Mitteilungsblatt der Bundesanstalt für Wasserbau, Karlsruhe. *(Measurements of flow in drainage channels with willow growth.)*

Florineth, F. (1982) Erfahrungen mit ingenieurbiologischen Maßnahmen bei Fließgewassern im Gebirge. Landschaftswasserbau, TU Wien, 3, 243–262. *(Experiences with bioengineered measures for watercourses in mountains.)*

Gilnrainer, G. (1986) Strukturierung von Stauräumen. 5. Sem. Landschaftswasserbau. TU Wien, 7, 271–306. *(Structuring of dam sites.)*

Gray, D.H. (1991) Preface. In: *Proceedings of Workshop on Biotechnical*

Stabilisation, 21–23 August, University of Michigan, Ann Arbor, Michigan.

Hartge, K.H. (1986) Bodenmechanische Probleme durch Dammbepflanzungen. 6. Sem. Landschaftswasserbau, TU Wien, 8, 17–34. *(Geotechnical problems resulting from planting on dams.)*

Horstmann, K. und Schiechtl, H.M. (1979) Künstliche Schaffung von Ökozellen. Garten und Landschaft, 3, 175–178, München. *(Artificial creation of ecological cells.)*

Karl, S. (1990) Erfahrungen mit der Uferbepflanzungen an Fließgewässern. In: 9. Sem. Landschaftswasserbau. TU Wien, 10, 427–454. *(Mulch layers – brush mat.)* *(Experiences with planting of banks on watercourses.)*

Kauch, E.P. (1992) Individuelle Lösungen für Fließgewässer oder Ausbaunormierungen. In: 12. Sem. Landschaftswasserbau, TU Wien, 13, 246–270. *(Individual solutions for watercourses or construction standardization?)*

Kelenc, H. (1986) Stauraumgestaltung an der Drau. In: 6. Sem. Landschaftswasserbau, TU Wien, 8, 135–148. *(Dam site formation at Drau.)*

Keller, E. (1937) Lebende Verbauung. Vorläufiger Bericht über den Werdegang praktischer Durchführungsversuche 2. Teil: Die bautechnische Anwendung und Durchführung der lebenden Verbauung. Wasserwirtschaft und Technik, Wien. *(Live construction. Preliminary Report on the development of practical completion 2. Part: The construction/technical application and completion of live-build construction.)*

Kruedener, A. (1951) Ingenieurbiologie. Verl. E. Reinhardt, München – Basel. *First publication specialising in bioengineering.) (Bioengineering.)*

Lange, G. und Lecher, K. (1989) Gewässerregelung, Gewässerpflege, 286 S., P. Parey Hamburg – Berlin. *(Water regulation and water maintenance.)*

Linke, H. (1964) Rasenmatten – ein Baustoff zur Ufersicherung. Wasserwirtschaft-Wassertechnik, 9, 269–270, Berlin. *(Turves – a building material for lake-shore stabilisation.)*

Meusel, H. (1965) Vergleichende Chorologie der Zentraleuropäischen Flora. Verl. Fischer, Jena. *(Plant species diversity.) (Comparative distribution of the Central European flora.)*

Nordin, A.R. (1993) Bioengineering to ecoengineering – Part One: the many names. In: *The International Group of Bioengineers*, Newsletter No. 3, December.

Pockberger, J. (1952) Der naturgemäße Wirtschaftswald. 136 S. Verl. G. Fromme, Wien. *(Nature appropriate forest management.)*

Prückner, R. (1965) Die Technik der Lebendverbauung. 200 S., Österr. Agrarverlag Wien. *(The technique of live-building).*

Schiechtl, H.M. (1973) Sicherungsarbeiten im Landschaftsbau. 244 S., Verl. G.D.W. Callwey, München. *(Fundamental reference book.)* *(Control works in landscape construction.)*

Schiechtl, H.M. (1992): Weiden in der Praxis; die Weiden Mitteleuropas, ihre Verwendung und ihre Bestimmung. 130 S., Verl. Patzer, Berlin – Hannover. *(Guidebook for willows suitable for bioengineering techniques in Middle Europe and the Alps.)* *(Willow practice; Central European willows, their use and designation.)*

Schiechtl, H.M. and Stern, R. (1996) *Ground Bioengineering Techniques for Slope Protection and Erosion Control.* Blackwell Science, Oxford.

Schütz, W. (1989): 25-jährige ingenieurbiologische Hangsicherungsmaßnahmen an der Brenner-Autobahn. Diplomarbeit,Univ. f. Bodenkultur, Wien, 212 S. *(Thesis on bioengineering development and maintenance.)* *(25-year-old bioengineering slope protection measures at the Brenner Autobahn.)*

Seidel, K. (1965) Phenolabbau im Wasser durch *Scirpus lacustris L.* während der Versuchsdauer von 31 Monaten. Naturwissenschaften, 52, 398 S. *(Phenol stripping water by* Scirpus lacustris L *during a 31 months trial period.)*

Seidel, K. (1968) Elimination von Schmutz- und Ballaststoffen aus belasteten Gewässern durch höhere Pflanzen. Zeitsch. Vitalstoffe-Zivilisations-krankheiten, 46 S. *(Elimination of waste and solid matter from polluted waters by higher plants.)*

Seidel, K. (1971) Wirkung höherer Pflanzen auf pathogene Keime in Gewässern. Naturwissenschaften 58, 150. *(Effects of higher plants on pathogen germs in water.)*

Seifert, A. (1941) Reise zu französichen Wasserstraßen. Deutsche Wasserwirtschaft, 8.

Seifert, A. (1965) Naturferner und naturnaher Wasserbau. Montana Verlag Zürich. *(Nature-near and nature-distant water construction.)*

Seifert, A. (1970) Bäume im Wasser. Garten und Landschaft. 5, 153–157, München. *(Trees in water.)*

Sortir, R.B. (1995) Soil bioengineering experiences in North America. In: *Vegetation and Slopes: Stabilisation, Protection and Ecology* (ed. D.H. Barker). Proceedings of Institution of Civil Engineers Conference, 29–30 September 1994. Thomas Telford, London.

Swiss Federal Institute of Water Management (1982) Hochwasserschutz an Fließgewässern; Wegleitung 1982, Eidgen. Druck- u. Materialzentrale, Bern. *(Flood protection of watercourses.)*

Tschermak, L. (1961) Wuchsgebietskarte des Österreichischen Waldes.

Forstl. Bundesvers.-Austalt Wien. *(Plant propagation – shrubs and small trees.)* *(Map of growing regions of the Austrian forest lands.)*

Vitek, E. (1986) Gestaltung und standortgerechte Bepflanzung von Dämmen. In: 6. Sem. Landschaftswasserbau TU Wien, 8, 1–16. *(Organisation of locally appropriate planting of dams.)*

Watschinger, E. und Dragogna, G. (1968) Problematica della diffesa del suolo: le sistemazioni elastiche. Monti e Boschi, XIX, 6, 5–15, Roma. *(Soil protection problems: flexible stabilisation.)*

Zeh, H. (1982) Verwendung von Geotextilien in der Ingenieurbiologie. Schweizer Baublatt Nr. 36. *(Use of geotextiles in bioengineering.)*

Further Reading

Temperate and general water bioengineering

Amoros, C., Roux, A.L., Reygrobellet, J.L., Bravard, J.P. and Pautou, G. (1987) A method for applied ecological studies of fluvial hydrosystems. *Regulated Rivers*, **1** 17–36.

Andrews, J. and Kinsman, D. (1990) *Gravel Pit Restoration for Wildlife*, RSPB, Sandy.

Armitage, P.D., Furse, M.T. and Wright, J.F. (1992) Environmental quality and biological assessment in British rivers – past and future perspectives. *Direccion de Investigacion y Formacion Agropesqueras: Gobierno Vasco.*

Ash, J.R.V. and Woodcock, E.P. (1988) The operational use of river corridor surveys in management. *Journal of the Institute of Water and Environment Management*, **2** 423–428.

Baines, C. and Smart, J. (1984) *A Guide to Habitat Creation.* Greater London Council.

Barnes, H.H. (1967) Roughness characteristics of natural channels. US *Geological Survey and Water Supply Paper*, **1849**.

Barrett, P.R.F. and Murphy, K.J. (1982) The use of diquat-algimate for weed control in flowing waters. *Proceedings EWRS 6th Symposium on Aquatic Weeds*, pp. 200–208.

Barrett, P.R.F. (1978) Aquatic weed control. Necessity and methods. *Fish Management*, **9**(3), 93–101.

Barrett, P.R.F. and Newman, J.R. (1993) The control of algae with barley straw. *PIRA Conference Proceedings: Straw – A Valuable Raw Material*, 20–22 April 1993.

Bayley, P.B. (1991) The flood pulse advantage and the restoration of river floodplain systems. *Regulated Rivers*, **6**(2) 75–86.

Bestman, L. (1983) Shore line – living construction material. *Wasser und Boden*, 35, 3, March. Verlag Paul Varey, Hamburg.

Bestman, L. (1984) Practical utilisation of living construction material and applied techniques. *Wasser und Boden*, 36, 1, January. Verlag Paul Varey, Hamburg.

Biggs, J. (1993) River Restoration Project: Summary of Phase 1, Report to RRP.

Boon, P.J., Calow, P. and Petts, G.E. (1992) *River Conservation and Management*. John Wiley, Chichester.

Brookes, A. (1981) *Waterways and Wetlands*. British Trust for Conservation Volunteers, Oxford.

Brookes, A. (1985) River channelisation, traditional engineering methods, physical consequences and alternative practices. *Progress in Physical Geography*, **9**, 44–73.

Brookes, A. (1988) *Channelised rivers: perspectives for environmental management*. John Wiley, Chichester.

Brookes, A. (1990) Restoration and enhancement of engineered river channels: some European experiences. *Regulated Rivers Research and Management*, **5**, 45–56.

Brookes, A. (1991) Geomorphology. In *River Projects and Conservation: A Manual for Holistic Appraisal* (ed. J.L. Gardiner), pp. 57–66. John Wiley, Chichester.

Brookes, A. (1992) Recovery and restoration of some engineered British river channels. In: *River Conservation and Management* (eds P.J. Boon, P. Calow and G.E. Petts) pp. 337–352. John Wiley, Chichester.

Brooks, A. and Agate, E. (1987) *Waterways and Wetlands*. British Trust for Conservation Volunteers, Oxford.

Buisson, R. and Williams, G. (1991) RSPB Action for lowland wet grasslands. *RSPB Conservation Review*, **5**, 60–64.

Chow, V.T. (1959) *Open Channel Hydraulics*. McGraw-Hill, New York.

Church, M. (1992) Channel morphology and typology. In: *The Rivers Handbook* (eds P. Calow and G.E. Petts), **1**, pp. 126–143. Blackwell, Oxford.

Construction Industry Research Information Association (1980) *Guide to the design of storage ponds for flood control in a partly urbanised catchment area*. CIRIA Technical Note 100.

Dawson, F.H. (1989) Ecology and Management of Water Plants in Lowland Streams. *FBA Annual Report*, pp. 13–60, Windermere.

Dawson, F.H., Griffiths, G.H. and Saunders, R.M.K. (1992) *River Corridor Strategic Overview Feasibility Study*. National Rivers Authority, Bristol.

Goldsmith, W. (1993) Lakeside Bioengineering. *Land and Water*, March/April.

Gordon, N.D., McMahon, T.A. and Finlayson, B.L. (1992) *Stream Hydrology. An introduction for Ecologists*. John Wiley, Chichester.

Gray, D.H. & Sotir, R.B. (1996) *Biotechnical and Soil Bioengineering Slope Stabilization: A Practical Guide for Erosion Control.* Wiley, New York.

Gregory, K.J. and Walling, D.E. (1973) *Drainage Basin Form and Process.* Edward Arnold, London.

Harrelson, C., Rawlins, C.L. and Potyondy, J.P. (1994) *Stream Channel: Reference Sites – An Illustrated Guide to Field Technique.* USDA Forest Service.

Haslam, S.M. (1981) *River Vegetation: its Identification, Assessment and Management. A Field Guide to the Macrophytic Vegetation of British Watercourses.* Cambridge University Press, Cambridge.

Hawkes, H.A. (1975) River Zonation and Classification. In: *River Ecology* (ed. B. Whitton), pp. 312–374. Blackwell, Oxford.

Hemphill, R.W. and Bramley, M.E. (1989) *Protection of River and Canal Banks.* Butterworths, London.

Holmes, N.T.H. (1983) *Classification of British rivers according to their flora. Focus on Nature Conservation* No. 3. Nature Conservancy Council, Peterborough.

Holmes, N.T.H. (1986) *Wildlife Surveys of Rivers in Relation to River Management.* A report to the Water Research Centre, Medmenham.

Holmes, N.T.H. (1993) River restoration as an integral part of river management in England and Wales. In: *Contributions to the European Workshop.* Ecological Rehabilitation of River Floodplains, Arnhem, The Netherlands, 22–24 September 1992. Report No. 11-6 under the auspices of the CHR/KHR, pp. 165–172.

Holmes, N.T.H. and Newbold, C. (1989) Nature conservation (A) – rivers as natural resources: and nature conservation (B). Sympathetic river management. In: *River engineering – part II, structures and coastal defence works.* Water No. 8 (ed. T.W. Brandon). IWGM, London.

Iversen, T.M., Kronvang, B., Madsen, B.L., Markmann and Nielsen, M.B. (1993) Re-establishment of Danish streams: restoration and maintenance measures. *Aquatic Conservation: Marine and Freshwater Ecosystems*, **3**, 73–92.

Junk, W.J., Bayley, P.B. and Sparks, R.E. (1989) The flood pulse concept in river-floodplain systems. In: *Proceedings of the International Large Rivers Symposium (LARS)*, Canadian Special Publication on Fisheries and Aquatic Sciences (ed. D.P. Dodge), **106**, pp. 110–127.

Kellerhalls, R., Church, M. and Bray, D.I. (1976) Classification and analysis of river processes. *American Society of Civil Engineers, Journal of the Hydraulics Division*, **102**, 813–829.

Kondolf, G.M. and Sales, M.J. (1985) Application of historical channel stability analysis to instream flow studies. *Proceedings of Symposium on Small Hydropower and Fisheries*, pp. 184–194, Colorado.

Kondolf, M.G. and Micheli, E.R. (1993) Evaluating success of stream restoration projects. *Journal of Environmental Management*.

Lane, E.W. (1955) The importance of fluvial geomorphology in hydraulic engineering. *ASCE Proceeding*, **81** (745).

Large, A.R.G. and Petts, G.E. (1992) Restoration of floodplaints: a UK perspective. In: *Contributions to the European Workshop*. Ecological Rehabilitation of River Floodplains, Arnhem, The Netherlands, 22–24 September 1992. Report No. 11–6 under the auspices of the CHR/KHR, pp. 173–180.

Ministry of Agriculture, Fisheries and Food, English Nature and The National Rivers Authority (1992) *Environmental Procedures for Inland Flood Defence Works*.

National Rivers Authority (1992) *A Guide to Bank Restoration and River Narrowing*. NRA Southern region.

National Rivers Authority (1992b) River corridor surveys; methods and procedures. *Conservation Technical Handbook* No. 1, NRA, Bristol.

National Rivers Authority (1993) *R and D Report on Bank Erosion on Navigable Waterways*. Projects 336, by Nottingham University (C.R. Thorne *et al.*). NRA, Bristol.

Nature Conservancy Council (1991) *Earth Science Conservation in Great Britain – A Strategy*. Nature Conservancy Council, Shrewsbury.

Nelson, A. and Nelson, K.D. (1973) *Dictionary of Water and Water Engineering*. Butterworths, London.

Newbold, C., Honnor, J. and Buckley, K. (1989) *Nature Conservation and the Management of Drainage Channels*. Nature Conservancy Council/ADA.

Newbold, C., Purseglove, J. and Holmes, N. (1983) *Nature Conservation and River Engineering*. Nature Conservancy Council, Shrewsbury.

Newson, M.D. (1992a) River conservation and catchment management; a UK perspective. In: *River Conservation and Management* (eds P.J. Boon, P., Calow and G.E. Petts) pp. 385–396. John Wiley, Chichester.

Newson, M.D. (1992b) *Land, Water and Development*. Routledge, London.

Nielsen, M.B. (1992) *Re-creation of Meanders and Other Examples of Stream Restoration in Southern Jutland, Denmark*. Senderlyllands Amt.

Nixon, M. (1966) Flood regulation and river training. In: *River Engineering and Water Conservation Works* (ed. R.B. Thorn).

Oplatka, M., Diez, C., Leuzinger, Y., Palmeri, F., Dibona, L. and Frossard, P. (1996) *Dictionary of Soil Bioengineering*. vdf Hochschulverlag AG.

Purseglove, J. (1988) *Taming the Flood – A History and Natural History of Rivers and Wetlands*. Oxford University Press/Channel 4 Television Company, Oxford.

Raven, P.J. (1986a) Changes of in-channel vegetation following two-stage channel construction on a small rural clay river. *Journal of Applied Ecology*, **23**, 333–345.

Raven, P.J. (1986b) Changes in waterside vegetation following two-stage channel construction on a small rural clay river. *Journal of Applied Ecology*, **23**, 989–1000.

Rheinalt, T. ap (1990) *River Corridor Surveys in Relation to the Nature Conservation Activities of the NRA*, National Rivers Authority, Bristol, report produced by Water Research Centre, Medmenham.

Richards, K.S. (1982) *Rivers; Form and Process in Alluvial Channels*. Methuen, London.

Richards, K.S. (ed.) (1987) *River Channels: Environment and Process*. Blackwell, Oxford.

Royal Society for the Protection of Birds (1978) *A Survey of the Birds of the River Severn*. RSPB, Sandy.

Royal Society for the Protection of Birds (1979) *A Survey of the Birds of the Avon*. RSPB, Sandy.

Royal Society for the Protection of Birds (1983) *Land Drainage and Birds in England and Wales: An Interim Report*. RSPB, Sandy.

Russell, H. (1992) Banking on Nature. *New Civil Engineer*, 19 November.

Sargent, G. (1991) *The importance of riverine habitats to bats in County Durham*. MSc thesis, University of Durham.

Schumm, S.A. (1977) *The Fluvial System*. Wiley Interscience, New York.

Sellin, R.J.H. and Giles, A. (1988) *A two stage channel flow*. Department of Civil Engineering, Contract Reference. IRAB/ENG/13/7/E.

Smith, H. and Drury, I. (1990) Otter Conservation in Practice. A report to the Water Research Centre.

Statzner, B., Gore, J.A. and Resh, V.H. (1988) Hydraulic stream ecology: observed patterns and potential applications. *Journal of North American Benthological Society*, **7**, 307–360.

United States Department of Agriculture: Soil Conservation Service (1992) *Soil Bioengineering for Upland Slope Protection and Erosion Reduction*. United States Department of Agriculture, Washington DC.

Walker, K.F., Thoms, M.C. and Sheldon, F. (1992) Effects of weirs on the littoral environment of the River Murray, South Australia. In: *River Conservation and Management* (eds P.J. Boon, P. Calow and G.E. Petts), pp. 271–292. John Wiley, Chichester.

Ward, D. (1992) (ed.) *Reedbeds for Wildlife*. Proceedings of a conference on creating and managing reedbeds with value to wildlife. RSPB/University of Bristol.

Ward, D., Holmes, N. and Jose, P. (1994) *The New Rivers and Wildlife Handbook*. RSPB/NRA/RSNC, Sandy.

Welcomme, R.L. (1992) River conservation: future prospect. In: *River Conservation and Management* (eds P.J. Boon, P. Callow and G.E. Petts) pp. 453–462. John Wiley, Chichester.

Wheeler, B. (1980) Plant communities in rich fen systems in England and Wales I. Introduction. Tall sedge and reed communities. *Journal of Ecology*, **68**, 365–395.

Williams, G., Newson, M. and Browne, D. (1988) Land drainage and birds in Northern Ireland. *RSPB Conservation Review*, **2**, Chapter 16.

Wright, J.F., Blackburn, J.H., Westlake, D.F., Furse, M.F. and Armitage, P.D. (1992) Anticipating the consequences of river management for the conservation of macroinvertebrates. In: *River Conservation and Management* (eds P.J. Boon, P. Callow and G.E. Petts) pp. 137–150. John Wiley, Chichester.

German language further reading

Anselm, R. (1990) Ingenieurbiologische Maßnahmen bei der Gewässerregulierung. In: 1. Sem. Landschaftswasserbau, TU Wien, 1, 70–104. *(Bioengineering measures for water regulation.)*

Baudirektion Des Kantons Bern (1989) Naturnahe Flachufer an Seen, 97 S., Bau Dion Bern. *(Bioengineering of shorelines of lakes.)*

Bayerisches Staatsministerium (1991) Flüsse, Bäche, Auen; pflegen und gestalten. 40 S. Oberste Baubehörde München. *(Rivers, streams and meadows: maintenance and design.)*

Begemann, W. (1971) Unweltschutz durch Gewässerpflege. 215 S., DRW Verl. Stuttgart. *(Environmental protection through water maintenance.)*

Begemann, W. und Schiechtl, H.M. (1986) Ingenieurbiologie. 216 S., Bauverlag, Wiesbaden – Berlin. *(Bioengineering.)*

Brandecker, H. (1991) Seeufergestaltung. Diplomarbeit, Univ. Bodenkultur Wien, 140 S. *(Lakeside formations.)*

Bretschneider, H. (1982) Taschenbuch der Wasserwirtschaft. 6. Aufl., P. Parey, Hamburg – Berlin. *(Handbook of water management.)*

Ehrengruber, C. (1989) Aspekte des naturnahen Flußbaues bei Buhnen. Diplomarbeit, Univ. für Bodenkultur, Wien. *(Aspects of water bioengineering groyne construction.)*

Göldi, Chr. und Walser, E. (1989) Wiederbelebungsprogramm für die Fließgewässer im Kanton Zürich. 38 S., Amt für Gewässerschutz, Zürich. *(Rejuvenation programme for watercourses in the Kanton of Zürich.)*

Habersack, H. (1990) Gewässerbetreuungskonzept Feistritz (Oststeiermark). Diplomarbeit Univ. Bodenkultur Wien, 328 S. mit Kartenteil. *(Watercourse maintenance concepts in Feistritz (Eastern Steiermark – Province or County.)*

Holzmann, H. (1985) Neuorientierung der Ziele im Wasserbau. In: Wasserbau-Entscheidung zwischen Natur und Korrektur. Akad. f. Naturschutz und Landschaftspflege, 2, 21–30, Laufen. *(New directions in the goals of water engineering.)*

Honsowitz, H. (1985) Die Abschätzung der Veränderung der hydraulischen Leistungsfähigkeit von revitalisierten Fließgewässerquerschnitten. In: 3. Sem. Landschaftswasserbau, TU Wien, 5, 307–350. *(Evaluation of changes in hydraulic capacity of revitalised watercourse cross-sections.)*

Jungwirth, M. (1980) Limnologische Aspekte naturbelassener und naturnahe verbauter Fließgewässer. In: Landschaftswasserbau, TU Wien, 1, 52–69. *(Limnological aspects of natural watercourses and bioengineering construction along watercourses.)*

Jungwirth, M. (1982) Ökologische Auswirkungen des Flußbaues. Wiener Mitt., Bd. 50, Kulturtechnik und Wasserwirtschaft heute, 4, 171–186, Wien. *(Ecological effects of river construction.)*

Karl, S. (1993) Landschaftsbau – Ausführung, Koordinierung und Kontrolle. In: 14. Sem. Landschaftswasserbau, TU Wien, 15, 349–360. *(Landscape construction, co-ordination and supervision.)*

Kirwald, E. (1955) Waldwirtschaft an Gewässern. 147 S. Wirtschafts- und Forstverl. Euting, Neuwied/Rh. *(Forest management along watercourses.)*

Kirwald, E. (1964) Gewässerpflege. 167 S., BLV München. *(Watercare.)*

Kirwald, E. (1982) Schäden und Nutzen von Gewässerwäldern. Jb. 1980 d. Ges. f. Ingenieurbiologie, 29–39, Verl. K. Krämer Stuttgart. *(Losses and gains from wooded watercourses.)*

Klötzli, F. (1980) Zur Verpflanzung von Streu- und Moorwiesen. ETH Zürich, Inst. f. Geobotanik, 5/80, 41–50. *(Transplantation of dry and water meadows.)*

Klötzli, F. (1981) Zur Reaktion verpflanzter Ökosysteme der Feuchtgebiete. Dat. Dok. Umweltschutz, 31, 107–117, Stuttgart. *(Aspects of transplanted ecology systems of wetlands.)*

Kröll, A. (1981) Die Stabilität von Steinschüttungen bei Sohlen- und Uferbefestigungen in Wasserströmungen. Inst. für Wasserwirtschaft und konstruktiven Wasserbau, TU Graz, Mitt. 23, 16–60. *(The stability of stone-dumping watercourse margins and bank protection.)*

Lautenschlager, E. (1989) Die Weiden der Schweiz. 136 S., Verl. Birkhäuser Basel. *(Vegetative propagated shrubs, species selection.) (The willows of Switzerland.)*

Mader, H. (1986) Revitalisierung Atlerbach/Salzburg. Diplomarbeit Univ. Bodenkultur Wien, 140 S. *(Revitalising Alterbach/Salzburg.)*

Martini, F. und Paiero, P. (1988) I Salici d'Italia. 160 S. Ed. Lint. Trieste. *(The willows of Italy.)*

Neumann, A. (1981) Die mitteleuropäischen Salixarten. 152 S., Mitt. Forstl. Bundesvers. Anstalt, 137, Wien. *(Plant diversity, species choice.) (Central European Salix species.)*

Pelikan, B. (1986) Revitaliseirung von Fließgewässern – ökologische Funktion wieder gefragt. Österr. Wasserwirtsch., 38, 3/4. Wien. *(Revitalisation of watercourses – ecological functions are in demand again.)*

Pretner, D. (1987) Die Rolle der Ingenieurbiologie im Flußbau der Steiermark. Diplomarb. Univ. Bodenkultur Wien, 136 S. *(The role of bioengineering in water construction along rivers in the Steiermark.)*

Rickert, K. (1990) Erfahrungen zur hydraulischen Charakterisierung von Gewässerquerschnitten. In: 9. Sem. Landschaftswasserbau, TU Wien, 10, 455–477. *(Experiences of hydraulic characterisation of cross-sections.)*

Rössler, J. (1989) Entwikclung von Röhrichtpflanzen bei der Renaturierung eines Fließgewässers. Zeitschr. f. Vegetationstechnik, 12, 34–42, München. *(Development of reeds in the renaturising of watercourses.)*

Schlüter, U. (1986) Pflanze als Baustoff. 332 S., Verl. Patzer, Berlin – Hannover. *(Application instruction book – outer alpine.) (Plants as building materials.)*

Schlüter, U. (1990) Laubgehölze – Ingenieurbiologische Einsatzmöglichkeiten. 164 S., Verl. Patzer, Berlin – Hannover. *(Species selection – lowlands – middle highland.) (Deciduous trees – bioengineered application possibilities.)*

Urstöger, F. (1984) Naturgemäßer Gewässerausbau. Diplomarbeit Univ. Bodenkultur Wien, 223 S. *(Nature appropriate water construction.)*

Verein für Ingenieurbiologie (1992): Bauen mit lebenden Pflanzen. 23 S., ETH Zürich, Inst. f. Kulturtechnik, 2, 92. *(Building with live plants.)*

Waltl, A. (1949) Der natürliche Wasserbau an Bächen und Flüssen. Amt Oberösterr. Landesreg., 3, 1–144, Linz. *(Natural water construction of streams and rivers.)*

Water Board of the Austrian Ministry of Agriculture and Forestry (1991) Schutzwasserbau, Gewässerbetreuung, Ökologie. 232 S., Wien. *(Protective water course construction, maintenance and ecology.)*

Wendelberger, E. (1986) Pflanzen der Feuchtgebiete. 223 S., BLV München. *(Plants in wetlands.)*

Willy, H. (1986) Vor- und Nachteile des naturnahen Gewässerlaufes im Vergleich zu kanalisierten Fließgewässern. Mitt. Inst. f. Wasserbau und Kulturtechnik, TU Karlsruhe, 195 S. *(Advantages and disadvantage of nature-near streams in canalised channels.)*

Wurzer, E. (1985) Natur- und landschaftsbezogener Schutzwasserbau – wesentliche Grundsätze des Leitfadens. In: 3. Sem. Landschaftswasserbau, TU Wien, 5, 1–16. *(Nature and landscape friendly water protection and water construction.)*

Zeh, H. (1983) Ingenieurbiologische Beispiele für das Bauen mit Pflanzen in der Schweiz. Garten und Landschaft, 6, 471–476, München. *(Construction techniques.) (Bioengineering examples for construction with plants in Switzerland.)*

Zeh, H. (1983) Ingenieurbiologische Bauweisen. 96 S., Bächtold AG, Bern. *(Bioengineering construction methods.)*

Zimmermann, A. und Otto, H. (1986) Konzept zur standortsgemäßen Bepflanzung regulierter Fluß- and Bachufer dür die Steiermark, Inst. f. Umweltwiss. und Naturschutz der Österr. Akademie der Wiss., 5/6, 5–57, Graz. *(Concept of locally appropriate planting of regulated stream and river banks in the Steiermark Region.)*

Index